The Beginner's Guide to Cloud Spotting

The Complete Spotter's Reference to Decoding the Clouds

Henry B. Preston

DISCLAIMER

The Beginner's Guide to Cloud Spotting: The Complete Spotter's Reference to Decoding the Clouds

First edition.

TABLE OF CONTENTS

PART 1

Unveiling the Mysteries of Clouds

CHAPTER 1

INTRODUCTION TO CLOUD SPOTTING

A HISTORY OF CLOUD OBSERVATION

Humans have looked to the sky for millennia, its ever-changing canvas a source of wonder, inspiration, and practical knowledge. From ancient proverbs predicting rain based on cloud formations to the sophisticated tools of modern meteorology, our relationship with clouds has been a fascinating journey of observation, classification, and scientific understanding.

The earliest documented observations of clouds likely coincided with the development of agriculture. Societies reliant on predictable weather patterns for crop success keenly observed the sky, noting how different cloud formations correlated with upcoming rain, fair weather, or storms. These observations were often passed down through generations in the form of proverbs, songs, and stories.

For example, a Chinese proverb dating back to 2000 BC states, "Mare's tails and mackerel scales, make sailors furl their sails." This proverb refers to the feathery cirrocumulus clouds, which often precede storms and strong winds. Similarly, ancient Greek and Roman texts contain references to cloud formations used for weather prediction. Aristotle, the influential Greek philosopher, devoted a section of his treatise "Meteorologica" (meaning "Discourse on Weather") to the study of clouds and their connection to weather patterns.

The scientific study of clouds took a significant leap forward in the early 19th century. Luke Howard, a pharmacist and amateur meteorologist in England, is credited with developing the first systematic cloud classification system. Frustrated by the lack of a standardized language for describing clouds, Howard published his groundbreaking work, "On the Modifications of Clouds" in 1803. He proposed a system based on Latin terms that categorized clouds based on their altitude and appearance. Howard identified three main cloud types:

- **Cirrus:** Feathery, wispy clouds composed primarily of ice crystals, found at high altitudes.
- **Stratus:** Sheet-like clouds that often obscure the entire sky, associated with drizzle or light rain.
- **Cumulus:** Puffy, dome-shaped clouds with flat bases, typically indicative of fair weather.

Howard's system, with some modifications, remains the foundation of modern cloud classification. Following Howard's work, other scientists like Jean-Baptiste Lamarck in France began to refine and expand upon this classification system.

The invention of photography in the mid-19th century revolutionized cloud observation. Previously, detailed descriptions were the only way to record cloud formations. Now, scientists could capture the intricate details of clouds with unprecedented accuracy. Pioneering meteorologists like Hildebrandsson in Sweden used photographs to create the first cloud atlases in the late 19th century. These atlases, featuring photographs alongside detailed descriptions, became invaluable tools for standardizing cloud identification and training future generations of meteorologists.

The 20th century saw further advancements in cloud observation with the invention of weather balloons and radar technology. Weather balloons allowed scientists to directly measure atmospheric conditions at different altitudes, providing valuable insights into cloud formation processes. Radar technology enabled meteorologists to "see" inside clouds, revealing their internal structure and composition. This newfound ability to study clouds from multiple angles and perspectives significantly improved our understanding of their role in weather systems.

Today, cloud observation remains a crucial aspect of meteorology. Ground-based observers equipped with standardized cloud charts and mobile apps contribute valuable data to weather forecasting models. Additionally, sophisticated satellite imaging provides real-time, global coverage of cloud formations, allowing meteorologists to track weather patterns and predict weather events with greater accuracy.

The history of cloud observation is a testament to our enduring fascination with the sky. From the proverbs of ancient cultures to the high-tech tools of modern science, our quest to understand clouds has not only improved our understanding of weather patterns but also fostered a deeper appreciation for the beauty and complexity of the natural world.

THE IMPORTANCE OF CLOUDS

Clouds, those ever-shifting formations drifting across the vast canvas of the sky, are far more than just visual spectacles. They play a critical role in shaping both our daily weather and the long-term health of our planet. Understanding these roles is key to appreciating the intricate dance between clouds, the atmosphere, and Earth's energy balance.

Regulating Sunshine and Precipitation

Sun Shields and Heat Traps: Clouds act as Earth's natural thermostat, regulating the amount of solar radiation reaching the surface and influencing temperatures. High, thin clouds like cirrus clouds allow most sunlight to pass through, warming the Earth's surface. Conversely, thick, low-level clouds like stratocumulus clouds act as a giant shade cloth, reflecting a significant portion of sunlight back into space, keeping the planet cool.

The Water Cycle in Action: Clouds are the linchpin of the water cycle, the continuous movement of water on, above, and below the Earth's surface. Water vapor, a gas invisible to the naked eye, rises from oceans, lakes, and land surfaces. As the air containing this vapor cools at higher altitudes, it condenses into tiny water droplets or ice crystals, forming clouds. When enough condensation occurs, these droplets or crystals become heavy enough to fall as precipitation – rain, snow, hail, or sleet – replenishing freshwater supplies and sustaining life on Earth.

Storm Signallers: Different cloud types can be telltale signs of upcoming weather events. Towering cumulonimbus clouds, for instance, with their anvil-shaped tops, often indicate thunderstorms brewing within, complete with lightning, heavy rain, and strong winds. Conversely, wispy cirrocumulus clouds, often arranged in a mackerel sky pattern, can signal fair weather ahead. By learning to identify different cloud types, you can become a weather detective, piecing together clues about the atmospheric conditions and predicting potential weather changes.

Earth's Delicate Balancing Act

The Greenhouse Effect: Clouds play a complex role in Earth's climate, impacting the greenhouse effect. Certain greenhouse gases, like carbon dioxide, methane, and water vapor, trap heat radiating from the Earth's surface, keeping our planet warm enough to sustain life. However, an excessive buildup of greenhouse gases can lead to a dangerous rise in global temperatures.

Reflectors and Insulators: High-altitude clouds, composed mainly of ice crystals, act like giant mirrors, reflecting a significant portion of incoming solar radiation back into space, mitigating the greenhouse effect. On the other hand, low-altitude clouds, particularly those laden with water droplets, can trap some of the Earth's outgoing heat radiation, amplifying the greenhouse effect.

A Delicate Balance: The net effect of clouds on climate depends on their type, altitude, and geographical location. High, thin cirrus clouds are generally considered to have a cooling effect, while low-level clouds with a higher water content can have a warming influence. Understanding this complex interplay between cloud types and their impact on the Earth's energy balance is crucial for our understanding of climate change.

Climate Change Concerns: The delicate balance between different cloud types and their influence on Earth's temperature is increasingly under threat due to climate change. Rising global temperatures can alter atmospheric circulation patterns, potentially leading to changes in cloud formation and distribution. For example, a decrease in high-altitude, reflective clouds could lead to a net warming effect. Additionally, changes in precipitation patterns could affect cloud cover in different regions, further impacting the global climate.

The Feedback Loop: Climate change and cloud cover can create a feedback loop. As global temperatures rise, more water vapor evaporates from oceans and land surfaces. This increased water vapor can lead to more cloud formation, potentially amplifying the warming effect. However, the exact nature of this feedback loop is complex and not fully understood by scientists.

Understanding the multifaceted role of clouds in shaping both weather and climate is essential for weather forecasting, climate change mitigation strategies, and for appreciating the intricate dance between Earth's atmosphere and its precious water resources. By becoming a cloud spotter, you not only gain a deeper appreciation for the beauty and artistry of the sky, but also contribute to our collective understanding of this critical component of our planet's well-being.

CLOUD SPOTTING AS A FUN AND ACCESSIBLE HOBBY

Gazing up at the sky has been a source of wonder and curiosity for humankind since the dawn of time. Yet, the ever-changing canvas of clouds often goes unnoticed, its beauty and complexity unappreciated. Cloud spotting, however, transforms this passive observation into an engaging and accessible hobby, enriching your daily walks, hikes, or even commutes with a sense of discovery and a newfound appreciation for the natural world.

The beauty of cloud spotting lies in its simplicity. Unlike many hobbies that require specialized equipment or extensive training, all you need is a curious mind and a willingness to look up. While a basic understanding of cloud types can enhance the experience, even a beginner can participate and find joy in the act of observing and appreciating the ever-changing formations above.

Here are just a few reasons why cloud spotting is a perfect hobby for anyone:

- **A Lifelong Learning Experience:** Clouds are a window into the fascinating world of atmospheric science. By learning to identify different cloud types, you'll gain a deeper understanding of weather patterns, how clouds form, and the role they play in our planet's climate. This knowledge can spark a lifelong fascination with the natural world and the complex processes that govern it.
- **A Stress-Buster and Mood Booster:** Taking a moment to appreciate the beauty of the clouds can be a powerful stress reliever. Stepping away from the daily grind and focusing on the vastness of the sky can induce a sense of calm and peace. The ever-changing cloud formations offer a visual feast, stimulating the imagination and fostering a sense of wonder. Studies have shown that spending time in nature, even simply looking at the sky, can have a positive impact on mental well-being and reduce stress levels.
- **A Gateway to Mindfulness:** Cloud spotting encourages mindfulness, a state of present-moment awareness. As you train your eye to identify different cloud formations, you become more attuned to the subtle changes in the sky, the interplay of light and shadow, and the movement of the clouds across the canvas. This mindful observation fosters a deeper connection with the natural world and a sense of appreciation for the present moment.
- **A Social Activity:** Cloud spotting is not a solitary pursuit. It can be a fun and engaging activity to share with friends, family, or fellow enthusiasts. Discuss the different cloud types you observe, share photos you capture, and learn from each other's experiences. Cloud spotting can be a great way to bond with loved ones and create lasting memories.
- **A Citizen Science Opportunity:** By learning to identify and record cloud observations, you can contribute to valuable citizen science projects. Many organizations collect cloud data from citizen scientists, which can be used for weather forecasting, climate research, and validating satellite imagery.

Here are some tips to get you started on your cloud spotting journey:

1. **Invest in a Cloud Identification Guide:** A good cloud identification guide will provide you with detailed descriptions and photographs of different cloud types, helping you learn to distinguish between them.
2. **Download a Cloud Spotting App:** Several mobile apps can help you identify clouds and learn about their associated weather patterns. These apps often include interactive features, quizzes, and cloud identification tools.
3. **Start Simple:** Don't overwhelm yourself with complex cloud classifications at the beginning. Begin by familiarizing yourself with the basic cloud types: high clouds (cirrus, cirrocumulus, cirrostratus), middle clouds (altocumulus, altostratus), low clouds (stratus, stratocumulus, nimbostratus), and vertical clouds (cumulus, cumulonimbus).
4. **Observe Throughout the Day:** Clouds can change dramatically throughout the day. Take some time each day, whether it's your morning walk or your evening commute, to look up and observe the sky. Pay attention to the shapes, colors, textures, and arrangements of the clouds.
5. **Join an Online Cloud Community:** There are numerous online forums and social media groups dedicated to cloud spotting. These communities offer a platform to share your observations, ask questions, and learn from experienced cloud spotters from around the world.
6. **Capture the Sky:** Don't forget to document your cloud spotting adventures! Take pictures of the interesting cloud formations you encounter. You might be surprised by the beauty you can capture with a simple smartphone camera.

Cloud spotting is a hobby that requires minimal investment but offers a wealth of rewards. With a little practice and curiosity, you can unlock the secrets hidden within the ever-changing canvas of the sky. So, the next time you find yourself outdoors, take a moment to look up. You might just discover a world of wonder and fascination waiting to be explored, one cloud at a time.

CHAPTER 2

UNDERSTANDING CLOUD FORMATION

ATMOSPHERIC CONDITIONS AND AIR PRESSURE

Clouds, those wispy or billowing formations adorning the sky, are not simply random puffs of water vapor. Their existence hinges on a complex interplay of atmospheric conditions, primarily temperature, pressure, and moisture. This section delves into the invisible playground where these conditions combine to create the stage for cloud formation.

Temperature's Crucial Role: Temperature plays a starring role in the drama of cloud formation. As air rises in the atmosphere, it cools due to decreasing pressure. This cooling process is crucial because warm air can hold more water vapor than cold air. As the rising air cools, it reaches a point of saturation, where it can no longer hold all the water vapor it contains. This excess vapor condenses around microscopic particles like dust or salt, forming tiny water droplets or ice crystals – the building blocks of clouds.

Understanding Adiabatic Cooling: The cooling of rising air is a process called adiabatic cooling. Imagine a parcel of air rising like a hot air balloon. As it ascends, it expands due to the decreasing pressure around it. However, this expansion happens without any exchange of heat with the surrounding environment. The energy used for this expansion comes from the internal energy of the air itself, causing its temperature to decrease. The rate of this cooling is determined by the type of air parcel – dry air cools at a faster rate than air saturated with water vapor (moist air). This difference in cooling rates is known as the lapse rate.

The Lapse Rate and Cloud Formation: The environmental lapse rate, the natural rate at which temperature decreases with altitude, plays a critical role in determining whether clouds will form. There are two main types of lapse rates to consider:

- **Environmental Lapse Rate (ELR):** This is the actual rate at which temperature decreases with altitude in a specific location.

- **Adiabatic Lapse Rate (ALR):** This is the rate at which a rising air parcel cools due to adiabatic cooling. There are two types of adiabatic lapse rates:
    - **Dry Adiabatic Lapse (DAL):** The rate at which dry air cools (approximately 9.8°C per kilometer).
    - **Moist Adiabatic Lapse (MAL):** The rate at which moist air cools (approximately 5°C per kilometer).

Cloud Formation and the Lapse Rate: Clouds form when the environmental lapse rate (ELR) becomes equal to or less than the adiabatic lapse rate (ALR) of the rising air parcel.

- **Condensation and Cloud Base:** When the ELR equals the DAL (for dry air), condensation occurs, marking the cloud base, the altitude where the rising air parcel becomes saturated.
- **Moist Air and Cloud Layers:** If the rising air is saturated with water vapor (moist air), the ELR needs to be equal to or less than the MAL (moist adiabatic lapse rate) for condensation to occur. This can happen at multiple levels, creating layered clouds.

Air Pressure's Influence: Air pressure, another key player in the atmospheric drama, is closely linked to temperature. As altitude increases, air pressure decreases. This decrease in pressure allows air parcels to rise and expand, promoting adiabatic cooling. High pressure systems, on the other hand, are generally associated with stable air masses and less cloud formation. Conversely, low pressure systems often trigger upward air movement and cloud formation due to the lower pressure at higher altitudes.

Moisture: While temperature and pressure create the stage for cloud formation, moisture is the essential ingredient. The amount of moisture present in the air determines how much water vapor can condense and form clouds. Areas with high humidity and abundant moisture content are more likely to see cloud formation compared to arid regions with low moisture levels.

Stability and Convection: The interplay of temperature, pressure, and moisture also influences atmospheric stability. Stable air masses resist vertical movement, leading to clear skies. Conversely, unstable air masses, where the ELR is less than the ALR of rising air parcels, promote convection – the rising of warm, less dense air. This upward movement leads to adiabatic cooling, condensation, and cloud formation.

Fronts and Cloud Formation: Fronts, boundaries between air masses with different temperatures and moisture content, play a significant role in cloud formation. As warm and cold air masses meet, differences in pressure and temperature create updrafts that promote condensation and cloud formation along the frontal boundaries. Different cloud types are often associated with specific types of fronts, allowing experienced cloud spotters to predict weather changes based on cloud formations.

Understanding these atmospheric conditions is the foundation for appreciating the intricate dance that leads to cloud formation. The next time you gaze skyward, remember the invisible playground where temperature, pressure, moisture, and air movement orchestrate the creation of those ever-changing formations above. By delving into this world, you'll gain a deeper appreciation for the beauty and complexity of clouds, not just as fleeting wisps in the sky, but as a testament to the dynamic processes that govern our atmosphere.

Additional Factors Affecting Cloud Formation

While temperature, pressure, moisture, and lapse rates are the primary drivers of cloud formation, several other factors can influence the type, shape, and size of clouds:

- **Wind Shear:** Wind shear, the change in wind speed or direction with altitude, can cause rising air parcels to be stretched or tilted. This stretching and tilting can enhance condensation and cloud formation, leading to the development of elongated or layered cloud structures.
- **Topography:** Mountains can act as giant barriers, forcing air to rise and cool adiabatically, leading to cloud formation on the windward side of the mountain. Conversely, the descending air on the leeward side (rain shadow) experiences warming due to compression, often resulting in clear skies.
- **Atmospheric Inversions:** Inversions are layers in the atmosphere where temperature increases with altitude, trapping cooler air

below. These inversions can inhibit convection and cloud formation. However, thin, sheet-like clouds like stratus can sometimes form at the base of inversions due to radiative cooling.

- **Stability and Lifting Mechanisms:** While atmospheric instability promotes cloud formation, specific lifting mechanisms are needed to initiate the rise of air parcels. These lifting mechanisms include:
    - **Convection:** As discussed earlier, rising warm air due to buoyancy can lead to cloud formation.
    - **Convergence:** When air masses converge horizontally, they are forced to rise, promoting condensation and cloud formation.
    - **Orographic Lifting:** As air is forced to rise over mountains, it cools and condenses, forming clouds.

Understanding Cloud Dynamics

Clouds are not static formations. They are constantly evolving, changing shape, size, and even type as they are buffeted by winds, influenced by temperature fluctuations, and affected by precipitation processes. Understanding these cloud dynamics is crucial for interpreting cloud formations and predicting potential weather changes.

- **Cloud Lifespan:** The lifespan of a cloud can vary greatly, depending on the type of cloud, atmospheric conditions, and wind patterns. Fair-weather cumulus clouds may last for as little as 20 minutes, while storm clouds associated with cumulonimbus formations can persist for several hours.
- **Precipitation and Dissipation:** As cloud droplets or ice crystals grow larger within a cloud, they eventually become heavy enough to fall as precipitation – rain, snow, hail, or sleet. This precipitation process removes water vapor from the cloud, leading to its eventual dissipation.
- **Cloud Seeding:** Cloud seeding is a technique used to artificially increase precipitation by introducing condensation nuclei (tiny particles) into clouds. This can be done by spraying silver iodide or dry ice into clouds under specific atmospheric conditions.

By delving into the world of atmospheric conditions, lapse rates, and cloud dynamics, you gain a deeper appreciation for the fascinating processes that orchestrate the formation, evolution, and eventual dissipation of clouds. The next time you look up at the sky, you'll see more than just wispy formations; you'll witness a dynamic theater where temperature, pressure, moisture, and wind play out their roles, shaping the ever-changing canvas above us.

HOW CLOUDS FORM IN THE SKY

Clouds, those wispy or billowing formations adorning the sky, wouldn't exist without a crucial first act: getting air to rise. While temperature, pressure, and moisture create the perfect recipe for cloud formation, a lifting mechanism is needed to kickstart the process. Imagine a pot of water on the stove – the water represents the air, and the heat source represents the factors promoting condensation.

But without bringing the pot to a boil (causing the water to rise), condensation wouldn't occur. Lifting mechanisms act like the invisible stovetop, forcing air parcels to ascend to cooler altitudes, triggering the chain reaction that leads to cloud formation.

This section delves into the key lifting mechanisms that propel air skyward, setting the stage for the condensation magic show that creates clouds.

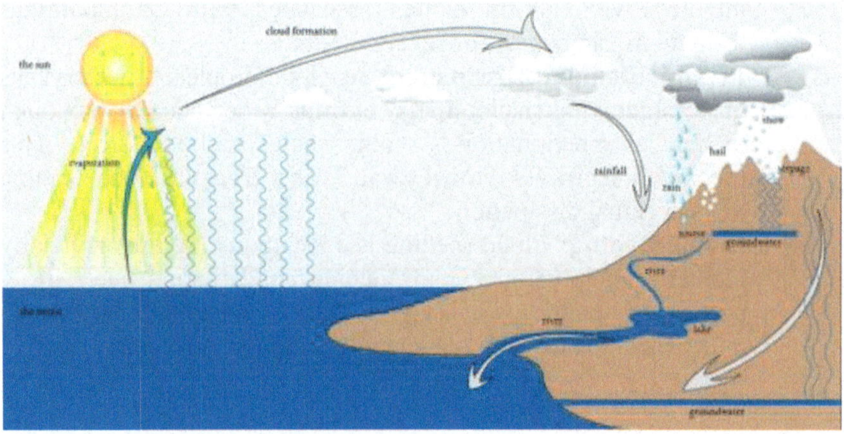

Convection

Convection, the rising of warm air due to buoyancy, is perhaps the most intuitive lifting mechanism. Imagine a hot air balloon – the heated air inside the balloon is less dense than the cooler surrounding air, causing it to rise. Similarly, on a sunny day, the Earth's surface absorbs solar radiation, warming the air in contact with it. This warm air expands and becomes less dense than the cooler air above. Due to buoyancy, the warm air parcel rises, cools adiabatically (explained in the previous section), and eventually reaches a point of saturation, forming a cloud. Fair-weather cumulus clouds are a classic example of clouds formed through convection.

Convergence

Convergence, the horizontal movement of air masses towards a common point, can also trigger cloud formation. Imagine two lanes of slow-moving traffic merging into one lane. As the air masses converge, they are forced upwards, leading to adiabatic cooling and potential cloud formation. Convergence zones are often associated with low-pressure systems, where air tends to sink towards the center. However, as this air converges near the center, it's forced to rise, creating lifting and promoting cloud formation.

Orographic Lifting

Mountains play a dramatic role in cloud formation through a process called orographic lifting. As air masses encounter a mountain barrier, they are forced to rise and flow over the obstacle. During this ascent, the air cools adiabatically, promoting condensation and cloud formation on the windward side (the side facing the approaching air) of the mountain. Conversely, the descending air on the leeward side (the downwind side) experiences warming due to compression, often resulting in clear skies or thinner, wispy clouds. This explains why mountainous regions often have dramatic cloud formations on one side and clear skies on the other.

Lifting by Fronts

Fronts, boundaries between air masses with contrasting temperatures and moisture content, are prime locations for cloud formation. As warm and cold air masses meet, the warm air is forced to rise over the denser, cooler air mass along a sloping frontal boundary. This rising warm air cools adiabatically, leading to condensation and cloud formation. Different types of fronts are associated with specific cloud formations, allowing experienced cloud spotters to use cloud patterns to predict weather changes. For example, thick, layered cloud formations often precede warm fronts, while towering cumulonimbus clouds can indicate the presence of a cold front.

Understanding Stability

The effectiveness of lifting mechanisms in creating clouds depends on atmospheric stability. Stable air masses resist vertical movement, and even if lifting occurs, the rising air parcel may not cool enough to reach saturation and form a cloud. Conversely, unstable air masses readily rise due to strong buoyancy or lifting mechanisms. As a result, unstable air is more likely to experience condensation and cloud formation.

Lifting Mechanisms Working Together

In reality, lifting mechanisms often work in concert. For example, convergence can occur near mountains, amplifying the lifting effect of orographic lifting. Similarly, frontal boundaries can be influenced by temperature differences caused by uneven heating of the Earth's surface, creating a complex interplay of factors that shape cloud formation.

Additional Lifting Mechanisms

While the mechanisms discussed above are the most common, other factors can contribute to lifting and cloud formation:

- **Upper-Level Divergence:** When air in the upper atmosphere diverges (spreads out horizontally), it can cause air below to rise, promoting cloud formation.

- **Turbulence:** Atmospheric turbulence caused by wind shear (change in wind speed or direction with altitude) can also create localized lifting and contribute to cloud formation.

Understanding lifting mechanisms is a key element in decoding the language of clouds. By recognizing the telltale signs of different lifting processes, you can gain valuable insights into atmospheric conditions and even predict potential weather changes.

CONDENSATION, PRECIPITATION, AND THE CLOUD LIFE CYCLE

Clouds, those ever-shifting formations gracing the sky, are not simply passive bystanders in the weather drama. They are active participants, playing a critical role in the water cycle – the continuous movement of water on, above, and below the Earth's surface. This section delves into the fascinating processes of condensation, precipitation, and the cloud life cycle, revealing how clouds transform invisible water vapor into rain, snow, hail, or sleet, and ultimately influence our weather patterns.

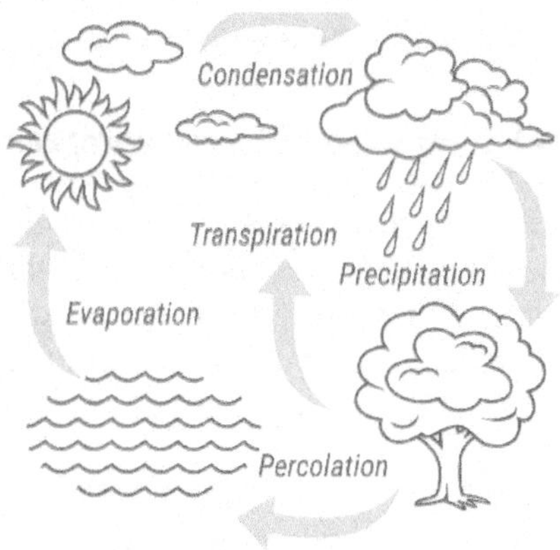

Condensation

The story begins with water vapor, a gas invisible to the naked eye. As discussed earlier, rising air parcels cool due to decreasing pressure. When the rising air reaches a point of saturation, it can no longer hold all the water vapor it contains. This crucial point marks the beginning of condensation – the transformation of water vapor from a gas to a liquid state.

The Role of Condensation Nuclei: However, condensation doesn't happen spontaneously. It requires microscopic particles suspended in the air called condensation nuclei. These particles, which can be dust, salt (especially near oceans), or even pollution particles, provide a surface for water vapor to condense upon. As water vapor molecules collide with these condensation nuclei, they begin to cluster together, forming tiny water droplets or ice crystals – the building blocks of clouds.

Temperature Dictates the Form: The temperature at the point of saturation determines whether the condensed water vapor forms liquid droplets or ice crystals. In warmer clouds with temperatures above freezing, condensation results in the formation of water droplets. Conversely, in colder clouds with temperatures below freezing, water vapor condenses directly into ice crystals. In some clouds, both water droplets and ice crystals can coexist, creating a mixed-phase cloud.

Precipitation

Condensation sets the stage for the next act – precipitation. As water droplets or ice crystals within a cloud grow larger and heavier, they eventually overcome the updrafts holding them aloft. This can happen through several mechanisms:

- **Collision-Coalescence:** When smaller droplets collide and merge with larger ones, they grow heavier and fall as rain. This process is particularly efficient in warm clouds containing a large number of small droplets.
- **The Bergeron-Findeisen Process:** In mixed-phase clouds, where both water droplets and ice crystals coexist, the ice crystals can grow at the expense of the water droplets through a process called the Bergeron-Findeisen process. As temperatures decrease with altitude, water vapor preferentially condenses onto the ice crystals due to their lower vapor pressure. This continuous growth of ice crystals eventually leads to precipitation, which can fall as snow, hail, or a combination of both.

The Journey to the Ground: The type of precipitation that reaches the ground depends on the temperature profile within the cloud and the atmosphere below.

- **Rain:** In warm clouds, where temperatures remain above freezing throughout the cloud depth, the falling water droplets reach the ground as rain.
- **Snow:** In cold clouds with temperatures below freezing throughout, the condensed water vapor forms ice crystals, which fall as snow.
- **Freezing Rain and Sleet:** If a layer of warm air exists above the ground, falling ice crystals can melt partially into rain droplets. However, if the temperature near the ground is again below freezing, these rain droplets refreeze into ice pellets known as sleet.
- **Hail:** Hail forms in strong thunderstorms with powerful updrafts that repeatedly lift and toss ice crystals within the cloud. As the ice crystals rise, they encounter supercooled water droplets (liquid water droplets at temperatures below freezing) which freeze onto their surface. This process continues, creating larger hailstones, which eventually become too heavy for the updrafts to support and fall to the ground.

The Cloud Life Cycle

Clouds are not static entities. They are constantly evolving, undergoing a life cycle that involves formation, growth, transformation, and dissipation. The specific details of this life cycle can vary depending on the type of cloud, atmospheric conditions, and the presence of precipitation.

Birth of a Cloud: The life cycle begins with lifting mechanisms that force air parcels to rise and cool adiabatically. As the air cools, condensation occurs, marking the formation of a cloud. Fair-weather cumulus clouds, often seen on sunny days, are a good example of this initial stage.

Growth and Transformation: As long as the lifting mechanism persists and sufficient moisture is available, a cloud can grow in size and thickness. Depending on the atmospheric conditions, the cloud may transform into a different type. For instance, small cumulus clouds may grow vertically into towering cumulonimbus clouds associated with thunderstorms.

Precipitation and Dissipation: When water droplets or ice crystals within a cloud grow large enough, they fall as precipitation. This precipitation process removes water vapor from the cloud, leading to its eventual dissipation. However, the story doesn't always end here. Depending on the wind patterns and ongoing atmospheric processes, several scenarios can unfold:

- **Partial Dissipation and Reformation:** If the lifting mechanism persists and sufficient moisture remains, a cloud may dissipate partially due to precipitation, but new cloud formations can arise within the same lifting zone, creating a cycle of formation and dissipation. This is often seen with afternoon thunderstorms, where rain showers can be followed by breaks in the clouds before new thunderstorm cells develop.
- **Dispersal by Winds:** Strong winds can shred and disperse clouds, particularly thin, high-level clouds like cirrus or cirrocumulus. As the cloud fragments are spread out, they become too thin for condensation to occur, leading to their complete disappearance.
- **Evaporative Lifting:** Even without precipitation, a cloud can dissipate through evaporation. When the surrounding air is drier than the air within the cloud, water vapor can evaporate from the cloud droplets or ice crystals, causing the cloud to shrink and eventually disappear. This is more common for thin, high-level clouds located in dry air masses.

Understanding the cloud life cycle not only helps you appreciate the dynamic nature of these formations but also provides clues about the weather patterns at play. For instance, the presence of towering cumulonimbus clouds with anvil-shaped tops suggests the possibility of thunderstorms, while the dissipation of a high-pressure system often leads to clear skies and fair weather.

Additional Factors Affecting Cloud Lifespan

Several additional factors can influence the lifespan of a cloud:

- **Stability:** Stable air masses generally support longer-lasting cloud formations compared to unstable air masses where updrafts can quickly lead to the growth and dissipation of clouds.
- **Temperature:** Warmer temperatures promote evaporation, leading to shorter cloud lifespans. Conversely, colder

temperatures can slow down evaporation processes, allowing clouds to persist for longer periods.

- **Humidity:** Higher humidity levels provide more moisture for cloud formation and growth, potentially extending their lifespan.

By understanding the interplay of condensation, precipitation, and the cloud life cycle, you gain valuable insights into the ever-changing drama unfolding in the sky above. The next time you gaze at a cloud formation, remember the fascinating processes at work – from the invisible transformation of vapor to the journey of water droplets or ice crystals, culminating in precipitation or dissipating into the vast canvas of the sky. With this knowledge, cloud spotting becomes more than just a passive pastime; it transforms into an engaging journey of discovery, where each cloud formation tells a captivating story about the weather and the dynamic processes that govern our planet.

CHAPTER 3

HOW CLOUDS AFFECT WEATHER

TYPES OF CLOUDS AND THEIR ASSOCIATED WEATHER PATTERNS

Gazing skyward at the ever-changing canvas of clouds is a source of wonder for many. But these formations hold more than just aesthetic beauty; they are messengers whispering tales of the weather brewing above. By learning to identify different cloud types and understanding their associated weather patterns, you can transform cloud spotting from a passive pastime into an engaging and informative hobby. This section serves as your field guide, equipping you to decode the language of the clouds and gain valuable insights into the atmospheric conditions playing out above.

Classifying the Clouds

Clouds are not a random assortment of wispy puffs. Meteorologists have classified them into distinct types based on their altitude, shape, and formation process. This classification system provides a universal language for understanding and discussing clouds around the world. Here's a breakdown of the major cloud types and their characteristics:

High-Level Clouds (5-13 km):

- **Cirrus (Ci):** These wispy, feathery clouds, often resembling strands of hair, are composed of ice crystals. Their presence generally indicates fair weather, although they can sometimes precede the arrival of a warm front and associated cirrostratus clouds.

- **Cirrocumulus (Cc):** Appearing as small white patches or ripples  resembling a mackerel sky, cirrocumulus clouds are also composed of ice crystals. They can signal fair or unsettled weather, depending on the presence of other cloud types.

- **Cirrostratus (Cs):** These thin, sheet-like clouds often create a  hazy or milky appearance across the entire sky. They can indicate fair weather or the approach of a warm front, and their presence can sometimes create halo phenomena around the sun or moon.

Mid-Level Clouds (2.5-6 km):

- **Altocumulus (Ac):** Characterized by patches, sheets, or rolls of  white or gray puffy elements, altocumulus clouds are composed of water droplets or a mixture of water droplets and ice crystals. Their presence can indicate fair weather or unsettled conditions, and they can sometimes precede the development of altostratus or nimbostratus clouds.
- **Altostratus (As):** These gray or bluish sheet-like clouds often  cover the entire sky, sometimes allowing the sun or moon to shine through faintly. They are composed of water droplets or a mixture of water droplets and ice crystals and often signify approaching precipitation, particularly light rain or snow.

Low-Level Clouds (Surface-2 km):

- **Stratus (St):** Appearing as a featureless gray layer, stratus clouds often obscure most of the sky. They are composed of tiny water droplets and are associated with drizzle, light rain, or snow. When stratus clouds touch the ground, they become fog.

- **Stratocumulus (Sc):** These gray or white puffy patches or rolls arranged in rows, waves, or even grids are composed of water droplets. They often indicate fair weather, although they can sometimes precede the development of nimbostratus clouds and associated precipitation.

- **Nimbostratus (Ns):** These dark, gray, featureless sheet clouds often produce continuous light or moderate rain or snow. They are composed of water droplets and ice crystals and are associated with widespread precipitation.

Vertical Development Clouds:

- **Cumulus (Cu):** These puffy, dome-shaped clouds with flat bases often resemble piles of cotton. They are composed of water droplets and are associated with fair weather, particularly during the afternoon when rising air currents can trigger their formation.

- **Cumulonimbus (Cb):** These towering clouds with anvil-shaped tops are the giants of the cloud kingdom. Composed of water droplets and ice crystals, they are associated with thunderstorms, heavy rain, hail, strong winds, and even tornadoes.

Here's a comparative illustration of the differences between the different types/classifications of clouds presented in tabular form:

Cloud Type	Altitude (km)	Appearance	Composition	Associated Weather
Cirrus (Ci)	5-13	Wispy, feathery strands	Ice crystals	Fair weather (may precede warm fronts)
Cirrocumulus (Cc)	5-13	Small white patches or ripples ("mackerel sky")	Ice crystals	Fair or unsettled weather
Cirrostratus (Cs)	5-13	Thin, sheet-like, hazy or milky appearance	Ice crystals	Fair weather (may precede warm fronts)
Altocumulus (Ac)	2.5-6	Patches, sheets, or rolls of white or gray puffy elements	Water droplets or mixed	Fair weather (may precede altostratus or nimbostratus)

Altostratus (As)	2.5-6	Gray or bluish sheet-like, covering entire sky	Water droplets or mixed	Approaching precipitation (light rain or snow)
Stratus (St)	Surface-2	Featureless gray layer, obscuring most of the sky	Tiny water droplets	Drizzle, light rain, or snow (fog at ground level)
Stratocumulus (Sc)	Surface-2	Gray or white puffy patches or rolls in rows, waves, or grids	Water droplets	Fair weather (may precede nimbostratus)
Nimbostratus (Ns)	Surface-2	Dark, featureless gray sheet	Water droplets and ice crystals	Continuous light or moderate rain or snow
Cumulus (Cu)	Surface-2	Puffy, dome-shaped with flat bases ("cotton wool")	Water droplets	Fair weather (particularly afternoon)
Cumulonimbus (Cb)	Surface-variable (can reach tropopause)	Towering giants with anvil-shaped tops	Water droplets and ice crystals	Thunderstorms, heavy rain, hail, strong winds, tornadoes

Linking Cloud Types to Weather Patterns

By identifying the cloud types present in the sky, you can gain valuable insights into the current and potential future weather patterns:

- **Fair Weather Signs:** The presence of high-level cirrus clouds, fair-weather cumulus clouds, or mid-level altocumulus clouds with a patchy appearance often indicates fair weather conditions, particularly in the short term.
- **Approaching Fronts:** The arrival of cirrostratus clouds can signal the approach of a warm front, while the thickening of altocumulus clouds and their transition into altostratus can indicate the approach of a cold front and potential precipitation.
- **Precipitation Warnings:** The presence of nimbostratus clouds or towering cumulonimbus clouds is a strong indicator of impending precipitation, ranging from light rain or snow to heavy downpours, thunderstorms, and hail.

Refining Your Cloud Spotting Skills

While understanding basic cloud types and their associated weather patterns provides a good foundation, becoming a more adept cloud spotter requires further exploration. Here are some additional tips to refine your skills:

- **Observe Cloud Features:** Look beyond the general type and pay attention to specific cloud features. For instance, the presence of virga (precipitation falling from a cloud that evaporates before reaching the ground) beneath altocumulus clouds can indicate the potential for afternoon showers. Similarly, the anvil shape of a cumulonimbus cloud can reveal the wind direction at high altitudes.
- **Consider Cloud Combinations:** The presence of multiple cloud types together can paint a more complete picture of the weather story. For example, scattered cumulus clouds beneath a layer of altostratus might suggest a transition towards overcast skies and potential drizzle.
- **Track Cloud Movement:** Observe the movement of clouds across the sky. Fast-moving cirrus clouds can indicate the

presence of jet streams and potential for unsettled weather. Conversely, slow-moving altocumulus clouds often suggest stable air and fair-weather conditions.

- **Local Knowledge is Power:** Familiarize yourself with the typical cloud patterns associated with different weather systems in your location. This local knowledge can help you refine your interpretations and make more accurate predictions based on the specific cloud formations you observe.
- **Consult Weather Resources:** Combine your cloud observations with weather forecasts from reputable sources like national meteorological agencies or weather apps. This integrated approach allows you to verify your interpretations and gain a more comprehensive understanding of the current and predicted weather patterns.

The Wider Significance of Clouds

Clouds play a crucial role beyond simply influencing our daily weather. They are vital players in Earth's climate system, acting as a giant blanket that regulates the planet's temperature. High-level cirrus clouds can trap heat radiating from the Earth's surface, contributing to a warming effect. Conversely, low-level stratus clouds can reflect incoming solar radiation back into space, creating a cooling effect. Understanding cloud types and their properties is essential for climate scientists studying these complex interactions and their impact on global temperatures.

Cloud spotting is a rewarding and lifelong pursuit. By delving deeper into cloud types, their associated weather patterns, and the broader role they play in our planet's climate system, you embark on a continuous learning journey. With each observation, you refine your skills, gain a deeper appreciation for the beauty and complexity of the sky, and transform cloud spotting from a casual pastime into a captivating exploration of the atmospheric world above.

PREDICTING THE SHORT-TERM FORECAST

Imagine gazing skyward and not just seeing wispy formations, but glimpsing the weather story unfolding above. By understanding cloud types and their associated weather patterns, you can transform from a passive observer to a keen interpreter of the celestial canvas. This section equips you with the knowledge to predict short-term weather changes, allowing you to make informed decisions and plan your day around the whims of the atmosphere.

Signs of Sunshine and Clear Skies

A bright and sunny day is often accompanied by the presence of certain cloud types:

- **Fair-Weather Cumulus (Cu):** These puffy white clouds with flat bases, often resembling cotton balls, are a classic sign of fair weather. They typically form during the afternoon due to rising warm air currents and dissipate as the sun sets and temperatures cool.
- **Cirrus (Ci):** These wispy, high-altitude clouds, sometimes described as "mares' tails," generally indicate fair or stable weather conditions. However, they can occasionally precede the arrival of a warm front, so keep an eye out for any thickening or increase in their coverage.
- **Altocumulus Patches (Ac):** When altocumulus clouds appear in scattered patches or ripples resembling a "mackerel sky," they often signal fair weather, particularly during the morning or evening hours. However, if these patches begin to merge and form a more extensive sheet, it could indicate a change towards overcast skies.

Clues for Incoming Clouds and Potential Rain

The presence of certain cloud types can signal the approach of moisture and possible precipitation:

- **Altostratus (As):** These gray or bluish sheet-like clouds can indicate approaching fair weather if they are thin and patchy. However, as they thicken and cover a larger portion of the sky, they often foreshadow the arrival of a warm front and associated light rain or drizzle.
- **Cirrostratus (Cs):** These thin, sheet-like clouds that create a hazy or milky appearance across the sky can signal fair weather.

However, they can also precede the arrival of a warm front, and their presence can sometimes create halo phenomena around the sun or moon.

- **Stratocumulus (Sc):** These gray or white puffy patches or rolls arranged in rows, waves, or even grids are often associated with fair weather, particularly during the morning or evening. However, if they thicken and become more extensive, they could indicate the approach of a low-pressure system and potential drizzle or light rain.

Signs of Potential Downpours and Thunderstorms

Certain cloud formations hold the telltale signs of impending storms:

- **Nimbostratus (Ns):** These dark, featureless gray sheet clouds often produce continuous light or moderate rain or snow. Their presence is a strong indicator of widespread precipitation, making them a reliable sign to grab an umbrella or adjust your outdoor plans.
- **Cumulonimbus (Cb):** These towering giants with anvil-shaped tops are the harbingers of thunderstorms. Composed of water droplets and ice crystals, they are associated with heavy rain, hail, strong winds, and even tornadoes. If you spot cumulonimbus clouds developing, it's wise to seek shelter indoors and be prepared for potentially severe weather.

Additional Factors for Short-Term Forecasting

While cloud identification forms the foundation for short-term weather prediction, other factors can contribute to a more comprehensive forecast:

- **Cloud Movement:** The speed and direction of cloud movement can provide valuable clues. Fast-moving cirrus clouds can indicate the presence of jet streams and potential for unsettled weather. Conversely, slow-moving altocumulus clouds often suggest stable air and fair weather conditions.

- **Cloud Thickness:** The thickness of a cloud layer can influence the amount of precipitation it might produce. Thin altostratus clouds might bring light drizzle, while thicker nimbostratus clouds could unleash a downpour.
- **Multi-Layered Clouds:** The presence of multiple cloud types stacked on top of each other can paint a more detailed picture. For example, altocumulus clouds above altostratus clouds might suggest a transition towards increased cloud cover and potential rain.

Don't rely solely on cloud observations for your short-term forecast. Integrate your cloud spotting skills with information from reputable weather sources like national meteorological agencies or weather apps. This combined approach allows you to verify your interpretations and gain a more comprehensive understanding of the current and predicted weather patterns.

Limitations and the Art of Interpretation:

It's important to remember that cloud-based weather prediction is not an exact science. Factors like local geography, microclimates, and sudden changes in atmospheric conditions can influence weather patterns. Here are some limitations to keep in mind:

- **Regional Variations:** Cloud-weather relationships can vary depending on your location. Familiarize yourself with the typical cloud formations associated with different weather systems in your area.
- **Time Scales:** Cloud observations are best suited for short-term forecasts, typically within the next few hours. Predicting weather patterns beyond that timeframe requires consulting meteorological data and forecasts.
- **Rapidly Evolving Conditions:** Weather can change quickly, especially with thunderstorms. While cumulonimbus clouds are a strong indicator of potential storms, their development and intensity can be difficult to predict precisely based solely on cloud observations.

Despite these limitations, cloud spotting offers a rewarding and dynamic way to engage with the weather. By observing cloud formations, movement, and interactions, you become a more informed participant in your environment.

Cloud spotting is not just about predicting the weather; it's about appreciating the beauty and complexity of the sky above. With each observation, you embark on a journey of discovery, unraveling the language of the clouds and their intricate dance with the atmosphere. As you hone your skills, you'll gain a deeper understanding of the natural world and transform a casual glance skyward into a captivating exploration of the ever-changing canvas above.

UNDERSTANDING CLOUD LAYERS

Clouds are not simply isolated entities. They exist within a stratified realm – the atmosphere – where temperature, pressure, and composition change dramatically with altitude. Understanding these atmospheric layers is crucial for comprehending cloud formation and the intricate processes that govern our weather patterns. This section serves as your guide, taking you on a journey through the invisible layers where clouds reside, revealing the unique characteristics that shape their appearance and behavior.

The Troposphere

Our journey begins in the lowest layer of the atmosphere, the troposphere. This vital layer, extending from the Earth's surface to an average height of around 10 kilometers (6.2 miles), is where all weather phenomena occur. Here, temperature generally decreases with increasing altitude, a critical factor in cloud formation. The troposphere also houses the majority of the atmosphere's water vapor, the essential ingredient for cloud development.

- **Cloud Residents:** The troposphere is home to a wide variety of cloud types, from the fluffy fair-weather cumulus clouds to the towering giants – cumulonimbus – associated with thunderstorms. The specific type of cloud that forms depends on the prevailing temperature, moisture content, and atmospheric processes at play.

- **Lifted Condensation Levels:** As air parcels rise within the troposphere due to various mechanisms like convection or convergence, they cool due to decreasing pressure. This cooling process, known as adiabatic cooling, eventually reaches a point where the rising air can no longer hold all the water vapor it contains. This critical point is called the lifting condensation level (LCL) – the altitude at which condensation occurs and clouds begin to form.
- **The Role of Condensation Nuclei:** Even when the LCL is reached, condensation doesn't happen spontaneously. Microscopic particles suspended in the air, called condensation nuclei, provide a surface for water vapor molecules to condense upon. These nuclei can be dust, salt (especially near oceans), or even pollution particles.

The Tropopause

The troposphere is capped by the tropopause, a transitional zone where the temperature gradient changes abruptly. The tropopause acts as a barrier, separating the troposphere's turbulent and dynamic weather systems from the more stable layers above. The specific height of the tropopause varies depending on factors like latitude and season, but it typically ranges from around 16 kilometers (10 miles) at the equator to 8 kilometers (5 miles) at the poles.

- **Limited Cloud Formation:** Due to the stable temperature profile and drier conditions, cloud formation is rare within the tropopause itself. However, wispy cirrus clouds, composed of ice crystals, can sometimes form at the fringes of the tropopause, particularly in the tropics.

The Stratosphere

Ascending beyond the tropopause, we enter the stratosphere, a vast expanse extending from roughly 15-50 kilometers (9.3-31 miles) above Earth's surface. This layer is characterized by a significant increase in temperature with altitude, a phenomenon caused by the absorption of ultraviolet (UV) radiation by ozone molecules.

- **The Ozone Layer:** The stratosphere houses the vital ozone layer, a region rich in ozone (O3) molecules. This layer acts as a protective shield, absorbing harmful ultraviolet radiation from the sun, preventing it from reaching the Earth's surface and safeguarding life as we know it.
- **Limited Cloud Formation:** Similar to the tropopause, cloud formation is uncommon within the stratosphere due to the extremely dry conditions. However, on rare occasions, special types of clouds, known as polar stratospheric clouds (PSCs), can form at very high altitudes within the polar stratosphere under extremely cold temperatures.

The Upper Atmosphere

The journey continues upwards, venturing beyond the stratosphere into the realm of the mesosphere, thermosphere, and exosphere. While these upper atmospheric layers play a crucial role in regulating Earth's temperature and protecting it from harmful solar radiation, they are not significant contributors to cloud formation due to their extremely low densities and temperatures.

Understanding Cloud Layers

By understanding the structure of the atmosphere and the unique characteristics of each layer, we gain valuable insights into cloud formation and the factors that influence their behavior.

- **Temperature and Cloud Types:** The temperature profile within the troposphere plays a critical role in determining the type of cloud that forms. Warmer temperatures favor the development of clouds composed of water droplets, while colder temperatures lead to the formation of ice crystals and clouds like cirrus or cirrostratus.
- **Stability and Cloud Formation:** The stability of the atmosphere also influences cloud formation. Stable air masses generally resist vertical movement, limiting cloud development. Conversely, unstable air masses promote rising air parcels, leading to adiabatic cooling and the formation of clouds. The presence of different cloud layers can also be indicative of atmospheric stability. For

instance, a well-defined layer of altostratus clouds often signifies a stable air mass with widespread lifting and condensation.

- **Precipitation and Cloud Layers:** The vertical extent of clouds within the troposphere can influence the type and amount of precipitation they produce. Deep, vertically developed clouds like cumulonimbus can generate heavy rain, hail, and even snow due to the extensive journey water droplets or ice crystals take within the cloud, allowing for collision-coalescence and growth. Conversely, thin, sheet-like clouds like stratus or altostratus might produce light drizzle or light snow due to the limited opportunity for growth within the cloud layer.

The Interconnectedness of the Atmosphere

Understanding cloud layers is not an isolated exercise. The atmosphere functions as a complex interconnected system, and the properties of each layer influence the behavior of others. Here are some key points to consider:

- **Energy Transfer:** The sun's energy drives atmospheric circulation patterns. Uneven heating of the Earth's surface leads to convection currents within the troposphere, causing rising air parcels and promoting cloud formation.
- **The Greenhouse Effect:** Greenhouse gases like carbon dioxide trap heat radiating from the Earth's surface, influencing the temperature profile of the atmosphere and potentially impacting cloud formation patterns.
- **Global Circulation:** Large-scale atmospheric circulation patterns, like jet streams, can influence the movement and development of cloud systems, shaping regional weather patterns.

Empowering Your Cloud Spotting:

Knowing the different cloud layers and their characteristics empowers you to become a more adept cloud spotter. Here are some tips to put your knowledge into action:

- **Identify Cloud Base Height:** By estimating the base height of a cloud, you can gain clues about the prevailing temperature and potential weather conditions. For instance, a low cloud base with cumulus clouds might suggest a warm and potentially humid environment, while a high cloud base with cirrus clouds could indicate colder temperatures at higher altitudes.
- **Consider Cloud Layers in Combination:** The presence of multiple cloud layers stacked on top of each other can paint a more complete picture of the atmospheric conditions. For instance, altocumulus clouds above altostratus clouds might suggest a transition towards increased cloud cover and potential rain, while cirrus clouds above clear skies might indicate the approach of a warm front.
- **Track Cloud Movement:** Observe the movement and interaction of clouds across different layers. Fast-moving cirrus clouds at high altitudes can indicate the presence of jet streams and potential for unsettled weather. Conversely, slow-moving altocumulus clouds within a stable layer might suggest fair weather conditions.

By delving deeper into the world of cloud layers, you unlock a new dimension in cloud spotting. The seemingly random formations in the sky transform into a story waiting to be unraveled, revealing the intricate interplay between temperature, pressure, moisture, and atmospheric dynamics that govern our weather patterns. With this knowledge, your gaze skyward transcends mere observation; it becomes a captivating exploration of the atmospheric dance that unfolds above, bringing you closer to understanding the ever-changing story of the clouds.

PART 2

Cloud Classifications in Detail

CHAPTER 4

HIGH CLOUDS

CIRRUS CLOUDS

Gazing skyward on a clear day, you might encounter wispy, delicate strands gracing the vast blue canvas. These ethereal formations are cirrus clouds, the highest and often the most captivating residents of the celestial realm. Often resembling feathers or wispy brushstrokes, cirrus clouds hold a unique place in the world of cloud spotting, offering clues about both fair weather and approaching weather systems. This section delves into the fascinating world of cirrus clouds, exploring their formation, characteristics, and the valuable insights they offer for understanding atmospheric conditions.

Cirrus clouds reside in the uppermost reaches of the troposphere, the lowest layer of the atmosphere where most weather phenomena occur. Their lofty domain, typically between 5 and 13 kilometers (3-8 miles) above Earth's surface, coincides with frigid temperatures ranging from -40°C to -60°C (-40°F to -76°F). Due to these extreme cold conditions, cirrus clouds are composed entirely of ice crystals, rather than water droplets that dominate clouds at lower altitudes.

Formation:

The formation of cirrus clouds hinges on a delicate balance between temperature and moisture. As air parcels rise within the troposphere, they cool due to decreasing pressure. When this rising air reaches a point where its temperature falls below the dew point temperature – the temperature at which condensation occurs – the water vapor present in the air condenses directly into ice crystals, bypassing the liquid water stage altogether. This process, known as sublimation, gives rise to the wispy, feathery structures that define cirrus clouds.

Appearance:

Cirrus clouds come in a variety of captivating forms, each adding a unique touch to the celestial canvas.

- **Horsetails (Cirrus uncinus):** These wispy strands often resemble the flowing tail of a horse, with a hooked or curved appearance at the end. Their presence can indicate strong winds blowing at high altitudes, sometimes associated with approaching jet streams.
- **Comma Clouds (Cirrus fibratus):** Resembling commas scattered across the sky, these clouds often form in bands or streaks and can signal fair or unsettled weather depending on other cloud formations present.
- **Mare's Tails (Cirrus spissatus):** These wispy, thread-like strands appear to stream across the sky, often likened to the flowing tail of a horse. While they are generally associated with fair weather, they can sometimes precede the arrival of a warm front, so keep an eye out for any thickening or increase in their coverage.

Fair Weather or Change on the Horizon?

Cirrus clouds are predominantly associated with fair weather conditions. Their presence often signifies stable atmospheric conditions with dry air at high altitudes. However, they can also offer subtle clues about potential changes in the weather patterns:

- **Thinning Cirrus:** Thinning cirrus clouds that gradually dissipate can indicate continued fair weather.
- **Thickening Cirrus:** A gradual thickening or increase in the coverage of cirrus clouds, particularly when accompanied by the appearance of cirrostratus clouds, can signal the approach of a warm front and potential for changes in wind direction and moisture levels.

The Broader Role of Cirrus Clouds

While cirrus clouds are primarily associated with fair weather, they play a vital role in Earth's climate system:

- **Greenhouse Effect:** Cirrus clouds, composed of ice crystals, can trap heat radiating from the Earth's surface, contributing to a warming effect. However, the net effect of cirrus clouds on climate is complex and depends on their thickness, altitude, and other factors.
- **Contrails:** Cirrus clouds can also form from contrails, the condensation trails left behind by high-flying aircraft. These contrails can persist for extended periods, potentially influencing cloud cover and impacting climate on a regional scale.

Cirrus clouds, with their delicate beauty and subtle messages, are a gateway to a deeper understanding of atmospheric processes. By observing and interpreting their characteristics, you embark on a journey that unveils the complex interplay between temperature, moisture, and air movement that governs our weather.

CIRROCUMULUS

Imagine gazing upwards on a bright day and encountering a captivating sight – a vast expanse of the blue sky adorned with a delicate, rippled pattern. These intriguing formations are cirrocumulus clouds, wispy patches or ripples composed of ice crystals that paint a unique picture in the celestial canvas. Often referred to as a "mackerel sky" due to their resemblance to fish scales, cirrocumutus clouds offer valuable clues about atmospheric conditions and can enhance your understanding of the weather patterns unfolding above. This section delves into the fascinating world of cirrocumulus clouds, exploring their formation, characteristics, and the insights they provide for interpreting the sky.

The Domain of Cirrocumulus

Cirrocumulus clouds reside in the upper reaches of the troposphere, sharing the lofty domain with cirrus clouds. They typically inhabit altitudes between 5 and 13 kilometers (3-8 miles) above Earth's surface, encountering frigid temperatures ranging from -40°C to -60°C (-40°F to -

76°F). Due to these extreme cold conditions, cirrocumulus clouds are composed entirely of ice crystals, much like their cirrus counterparts.

Formation

The formation of cirrocumulus clouds hinges on the interplay between rising air parcels, saturation, and the presence of ice nuclei. As air ascends within the troposphere due to convection or convergence, it cools due to decreasing pressure. When this rising air reaches a level where its temperature dips below the frost point – the temperature at which water vapor sublimates directly into ice crystals – saturation occurs. However, unlike cirrus clouds that form from widespread lifting, cirrocumulus tend to develop in patchier formations due to the sporadic presence of ice nuclei, microscopic particles that provide a surface for ice crystals to form.

Appearance

Cirrocumulus clouds come in a variety of captivating forms, adding a touch of intricate detail to the sky:

- **Patchy Sheets:** These formations appear as scattered white patches resembling scales or ripples across a blue canvas. They are the most common type of cirrocumulus cloud and often signify fair weather.
- **Ripple Patterns:** These clouds exhibit a wave-like or rippled appearance, often likened to a "mackerel sky." This pattern is caused by atmospheric waves propagating through the upper troposphere, and their presence can indicate fair or unsettled weather depending on other factors.
- **Lacunae:** Occasionally, gaps or holes may appear within a sheet of cirrocumulus clouds, known as lacunae. These gaps can be caused by sinking air parcels within the cloud layer or by variations in the distribution of ice nuclei.

Fair Weather or Something More?

Cirrocumulus clouds are generally associated with fair weather conditions. Their presence often indicates stable atmospheric conditions with dry air at high altitudes. However, depending on their characteristics and the presence of other cloud formations, they can offer additional insights:

- **Thin Patches:** Thin, scattered patches of cirrocumulus in a predominantly blue sky suggest continued fair weather.
- **Thickening Patches:** A gradual thickening or increase in the coverage of cirrocumulus clouds, particularly when accompanied by the appearance of altocumulus clouds, can indicate the approach of a warm front and potential changes in wind direction and moisture levels.
- **Widespread Ripples:** Extensive, wave-like cirrocumulus formations, particularly when accompanied by cirrus clouds, can suggest the presence of jet streams and potential for unsettled weather patterns.

Enhancing Your Cirrocumulus Spotting Skills

Here are some tips to become a more adept cirrocumulus cloud spotter:

- **Observe Cloud Texture:** Pay close attention to the texture of cirrocumulus clouds. Thin, patchy formations generally indicate fair weather, while thicker sheets or extensive ripple patterns might suggest a change in weather conditions.

- **Identify Accompanying Clouds:** Cirrocumulus clouds rarely appear in isolation. Observe the presence of other cloud types, particularly at lower altitudes. For instance, cirrocumulus preceding a layer of altocumulus might signal a thickening mid-level cloud layer and potential for showers, while cirrocumulus coexisting with clear skies suggests continued fair weather.
- **Track Cloud Movement:** Monitor the movement of cirrocumulus clouds. Fast-moving cirrocumulus, particularly when arranged in wave-like patterns, can indicate the presence of jet streams high in the atmosphere and potential for changes in wind direction. Conversely, slow-moving cirrocumulus suggests stable air and fair weather conditions.

By observing and interpreting the characteristics of cirrocumulus clouds, you gain valuable insights into the dynamic processes unfolding within the upper troposphere. These wispy formations, often resembling a "mackerel sky," can reveal information about atmospheric stability, the presence of jet streams, and potential changes in weather patterns.

CIRROSTRATUS

When you look skyward on a clear day, you might encounter a subtle transformation – a gentle softening of the vibrant blue canvas. This hazy veil, often barely perceptible, is the domain of cirrostratus clouds. These wispy, sheet-like formations, composed entirely of ice crystals, hold a unique place in the world of cloud spotting. While seemingly innocuous, cirrostratus clouds offer valuable insights into atmospheric conditions and can unveil the presence of subtle shifts in the weather patterns. This section delves into the fascinating world of cirrostratus clouds, exploring their formation, characteristics, and the stories they reveal about the sky above.

The Realm of Cirrostratus

Cirrostratus clouds reside in the uppermost reaches of the troposphere, sharing the lofty domain with cirrus and cirrocumulus clouds. They typically inhabit altitudes between 5 and 13 kilometers (3-8 miles) above Earth's surface, encountering frigid temperatures ranging from -40°C to -

60°C (-40°F to -76°F). Due to these extreme cold conditions, cirrostratus clouds are composed entirely of ice crystals, much like their cirrus counterparts.

Formation

The formation of cirrostratus clouds hinges on a large-scale lifting of moist air within the upper troposphere. As air parcels ascend due to processes like convergence or large-scale weather systems, they cool adiabatically (cool due to decreasing pressure). When this rising air reaches a level where its temperature dips below the frost point, water vapor sublimates directly into ice crystals, forming a widespread sheet of cirrostratus clouds. This sheet-like structure differentiates cirrostratus from cirrocumulus clouds, which exhibit patchier formations due to the sporadic distribution of ice nuclei.

Appearance

Cirrostratus clouds are known for their subtle and ethereal appearance. Unlike their more defined cirrus counterparts, they don't exhibit distinct wispy strands or dramatic formations. Instead, they manifest as a hazy, veil-like layer that can obscure the sharpness of the blue sky, sometimes giving it a milky or whitish appearance. Here's a closer look at some key characteristics:

- **Thin Sheet:** Cirrostratus clouds typically appear as a thin, sheet-like layer that can cover a large portion of the sky. They often reduce the sun's intensity without completely obscuring it, casting a weak shadow.
- **Halo Phenomena:** One of the most distinctive characteristics of cirrostratus clouds is their association with halo phenomena. When sunlight or moonlight refracts through the ice crystals in cirrostratus clouds, it can produce a dazzling display of colored rings or arcs around the sun or moon, known as halos. The specific type of halo observed depends on the size and orientation of the ice crystals within the cloud layer.
- **Sun Halo vs. Moon Halo:** Halos around the sun tend to be white or faintly colored, while those around the moon appear predominantly white due to the weaker moonlight.

Fair Weather or Change on the Horizon?

Cirrostratus clouds can be associated with both fair weather and approaching weather systems:

- **Fair Weather:** Thin cirrostratus clouds, particularly when accompanied by clear skies or cirrocumulus, can signify fair weather conditions. Their presence suggests stable atmospheric conditions with dry air at high altitudes.
- **Approaching Warm Front:** A gradual thickening of cirrostratus clouds, sometimes accompanied by the disappearance of halos, can indicate the approach of a warm front. As the warm front advances, the cirrostratus layer thickens, and the ice crystals within the cloud may begin to melt, reducing the distinctness of halos. In some cases, this thickening cirrostratus can transition into altostratus clouds as the warm front progresses.

Enhancing Your Cirrostratus Spotting Skills

Here are some tips to become a more adept cirrostratus cloud spotter:

- **Observe Transparency:** Pay attention to the transparency of the sky. Thin cirrostratus clouds might only create a subtle hazy effect, while thicker layers can noticeably reduce the clarity and vibrancy of the blue sky.
- **Look for Halos:** Cirrostratus clouds are the primary cloud type associated with halo phenomena. If you observe a halo around the sun or moon, it's a strong indicator of the presence of cirrostratus clouds. Be sure to note the halo's color and shape, as these characteristics can provide clues about the size and orientation of the ice crystals within the cloud layer.
- **Monitor Changes in Cloud Cover:** Track the changes in cirrostratus cloud cover over time. A gradual thickening of the cirrostratus layer, particularly when accompanied by the disappearance of halos, could signal the approach of a warm front and potential changes in weather patterns.

By observing and interpreting the characteristics of cirrostratus clouds, you gain valuable insights into the subtle processes unfolding within the upper troposphere. Their presence can indicate stable atmospheric conditions, the approach of weather systems, and even offer clues about the radiative balance affecting Earth's climate.

CHAPTER 5

MIDDLE CLOUDS

ALTOCUMULUS

When you look skyward on a bright day, you might encounter a captivating sight – a vast expanse of blue adorned with patches of white puffs, resembling a field of fluffy cotton balls. These formations are altocumulus clouds, residents of the mid-altitude realm. Often appearing in sheets, rolls, or patches, altocumulus clouds offer a glimpse into the atmospheric conditions of the middle troposphere and can hint at potential changes on the horizon. This section delves into the fascinating world of altocumulus clouds, exploring their formation, characteristics, and the valuable clues they provide for interpreting the sky.

The Domain of Altocumulus

Altocumulus clouds reside in the middle layer of the troposphere, occupying a space between the high-altitude cirrus and cirrocumulus clouds and the lower-level stratocumulus and cumulus clouds. They typically inhabit altitudes between 2 and 6 kilometers (1.2-3.7 miles) above Earth's surface, encountering temperatures ranging from -15°C to -35°C (5°F to -31°F). Due to these moderate temperatures, altocumulus clouds can be composed of a mixture of water droplets and ice crystals, depending on the specific atmospheric conditions.

Formation

The formation of altocumulus clouds hinges on a balance between rising air parcels, saturation, and the presence of condensation nuclei. As air ascends within the troposphere due to convection or convergence, it cools adiabatically (cools due to decreasing pressure). When this rising air reaches a level where its temperature dips below the dew point temperature for water droplets (or the frost point temperature for ice crystals), condensation or sublimation occurs, leading to the formation of altocumulus clouds. The presence of condensation nuclei, microscopic particles that provide a surface for water vapor to condense upon, plays a crucial role in this process.

Appearance

Altocumulus clouds come in a variety of captivating forms, each adding a unique touch to the celestial canvas:

- **Patchy Sheets:** These formations appear as scattered white patches resembling a flock of sheep or a field of cotton balls. They are the most common type of altocumulus cloud and often signify fair weather conditions.
- **Rolls and Waves:** Altocumulus clouds can also exhibit a wave-like or rippled appearance, often likened to mackerel scales or rows of sausages. These patterns are caused by atmospheric waves propagating through the middle troposphere, and their presence can indicate fair or unsettled weather depending on other factors.
- **Lacunae:** Occasionally, gaps or holes may appear within a sheet of altocumulus clouds, known as lacunae. These gaps can be caused by sinking air parcels within the cloud layer or by variations in the distribution of condensation nuclei.

Fair Weather or Something More?

Altocumulus clouds are generally associated with fair weather conditions. Their presence often indicates stable atmospheric conditions with dry air at mid-altitudes. However, depending on their characteristics and the presence of other cloud formations, they can offer additional insights:

- **Thin Patches:** Thin, scattered patches of altocumulus in a predominantly blue sky suggest continued fair weather.
- **Thickening Patches:** A gradual thickening or increase in the coverage of altocumulus clouds, particularly when accompanied by the appearance of cirrostratus clouds, can indicate the approach of a warm front and potential changes in wind direction and moisture levels.
- **Patchy Sheets with Shading:** Altocumulus sheets with some patches appearing darker or more shaded than others can indicate the presence of convective activity within the cloud layer. These darker patches may signify the development of towering cumulus clouds in the lower atmosphere, potentially leading to showers or thunderstorms.

Enhancing Your Altocumulus Spotting Skills

Here are some tips to become a more adept altocumulus cloud spotter:

- **Observe Cloud Texture:** Pay close attention to the texture of altocumulus clouds. Thin, patchy formations with a cotton-ball-like appearance generally indicate fair weather, while thicker sheets or extensive wave patterns might suggest a change in weather conditions.
- **Identify Accompanying Clouds:** Altocumulus clouds rarely appear in isolation. Observe the presence of other cloud types, particularly at higher and lower altitudes. For instance, altocumulus preceding a layer of cirrostratus could signal the approach of a warm front, while altocumulus coexisting with towering cumulus clouds below might suggest potential for showers or thunderstorms.
- **Track Cloud Movement:** Monitor the movement of altocumulus clouds. Fast-moving altocumulus, particularly when arranged in wave-like patterns, can indicate the presence of jet streams in the upper troposphere and potential for changes in wind direction. Conversely, slow-moving altocumulus suggests stable air and fair weather conditions.

ALTOSTRATUS

When you look skyward on a day that seems oddly muted, the vibrant blue canvas might appear veiled by a thin, featureless sheet. This hazy obscurity often signifies the presence of altostratus clouds, residents of the middle troposphere. While they may lack the dramatic formations of their towering cumulus cousins or the wispy elegance of cirrus, altostratus clouds play a crucial role in the celestial story, offering valuable clues about atmospheric conditions and potential weather developments. This section delves into the fascinating world of altostratus clouds, exploring their formation, characteristics, and the insights they provide for interpreting the sky.

The Realm of Altostratus

Altostratus clouds reside in the middle layer of the troposphere, sharing the airspace with altocumulus clouds. They typically inhabit altitudes between 2 and 6 kilometers (1.2-3.7 miles) above Earth's surface, encountering temperatures ranging from -15°C to -35°C (5°F to -31°F). Due to these moderate temperatures, altostratus clouds can be composed of a mixture of water droplets and ice crystals, depending on the specific atmospheric conditions.

Formation: A Widespread Sheet of Water and Ice

The formation of altostratus clouds hinges on large-scale lifting of moist air within the middle troposphere. As air ascends due to processes like convergence or large-scale weather systems, it cools adiabatically (cools due to decreasing pressure). When this rising air reaches a level where its temperature dips below the dew point temperature for water droplets (or the frost point temperature for ice crystals), widespread condensation or sublimation occurs, leading to the formation of a vast sheet of altostratus clouds. Unlike altocumulus clouds that exhibit patchy formations due to the uneven distribution of condensation nuclei, altostratus clouds develop a more uniform sheet-like structure due to the extensive lifting and condensation processes involved.

Appearance

Altostratus clouds are known for their subtle yet impactful presence in the sky. Here's a closer look at their key characteristics:

- **Sheet-Like Structure:** Altostratus clouds appear as a featureless, grayish-white sheet that can obscure a significant portion of the sky. They often diffuse the sunlight, casting a hazy veil over the landscape and reducing the overall clarity of the blue sky.
- **Sun and Moon Visibility:** Altostratus clouds are thin enough to allow the sun or moon to shine through, but their presence dims the light source, creating a weak and often watery halo around it.
- **Virga and Precipitation:** In some cases, precipitation may fall from altostratus clouds in the form of virga – wispy trails of rain or snow that evaporate before reaching the ground. However, if the altostratus layer thickens sufficiently, the virga can transition into light rain or snow at the surface.

Fair Weather or Warning Sign?

Altostratus clouds can be associated with both fair weather and approaching weather systems:

- **Fair Weather:** Thin altostratus sheets, particularly when accompanied by clear skies or patchy altocumulus, can signify fair weather conditions. Their presence suggests stable atmospheric conditions with dry air at mid-altitudes.

- **Approaching Warm Front:** A gradual thickening of the altostratus layer, sometimes accompanied by the disappearance of the sun's halo, can indicate the approach of a warm front. As the warm front advances, the altostratus layer thickens, and the water droplets within the cloud may begin to warm, leading to the dissipation of the halo. In some cases, this thickening altostratus can transition into nimbostratus clouds as the warm front progresses, bringing widespread precipitation.
- **Approaching Occluded Front:** The presence of a thick altostratus layer, along with falling precipitation at the surface, can suggest the presence of an occluded front – a complex weather system formed by the merging of a warm front and a cold front.

Enhancing Your Altostratus Spotting Skills

Here are some tips to become a more adept altostratus cloud spotter:

- **Observe Transparency:** Pay attention to the overall transparency of the sky. Thin altostratus sheets might only create a subtle hazy effect, while thicker layers can significantly reduce visibility and obscure the sun or moon to the point where only a faint outline remains.
- **Look for Halos:** While less prominent than with cirrostratus clouds, altostratus clouds can sometimes produce weak halos around the sun or moon. If you observe a faint, watery halo surrounding the sun or moon, it could be an indicator of altostratus clouds.
- **Monitor Cloud Changes:** Track the changes in the altostratus cloud layer over time. A gradual thickening of the sheet, particularly when accompanied by the disappearance of the halo and the presence of virga, could signal the approach of a warm or occluded front and potential changes in weather patterns.

By observing and interpreting the characteristics of altostratus clouds, you gain valuable insights into the subtle processes unfolding within the middle troposphere. Their presence can indicate stable atmospheric conditions, the approach of weather systems, and even offer clues about the radiative balance affecting Earth's climate.

CHAPTER 6

LOW CLOUDS

STRATUS

When you look skyward on a gloomy day, a vast sheet of gray might dominate the scene, casting a somber mood over the landscape. This ubiquitous blanket of cloud is likely stratus, the lowest layer in the cloud classification system. Often associated with drizzle and dreary weather, stratus clouds play a vital role in Earth's atmospheric processes. This section delves into the world of stratus, exploring their formation, characteristics, and the stories they reveal about the conditions below.

The Realm of Stratus

Stratus clouds reside in the lowest level of the atmosphere, the troposphere. They typically inhabit altitudes between 0 and 1.5 kilometers (0-1 mile) above Earth's surface, encountering temperatures ranging from -5°C to 15°C (23°F to 59°F). Due to these mild temperatures, stratus clouds are composed entirely of water droplets, unlike their high-altitude counterparts that may contain ice crystals. In some instances, stratus clouds can form in contact with the ground, transitioning into fog.

Formation

The formation of stratus clouds hinges on the cooling and saturation of air near Earth's surface. This cooling can occur through several mechanisms:

- **Radiational Cooling:** On clear nights, as the ground loses heat through radiation, the air in contact with it cools as well. If this cooling is sufficient, the air temperature can dip below the dew point, leading to condensation and the formation of stratus clouds.
- **Adiabatic Cooling:** As air ascends gentle slopes or over bodies of water due to processes like convection or forced lifting, it cools adiabatically (cools due to decreasing pressure). When this rising air reaches a level where its temperature falls below the dew point, condensation occurs, and stratus clouds form.
- **Advection:** The horizontal movement of cool, moist air masses over warmer surfaces can also trigger stratus cloud formation as the air cools due to contact with the underlying surface.

Appearance

Stratus clouds are known for their uniform, featureless appearance, often resembling a vast, gray blanket obscuring the sky. Here's a closer look at their key characteristics:

- **Sheet-Like Structure:** Stratus clouds appear as a featureless, grayish sheet that can cover a large portion of the sky. They often diffuse sunlight, casting a dull and gloomy light over the landscape.
- **Sun and Moon Visibility:** Stratus clouds are thick enough to obscure the sun or moon completely, reducing visibility and creating a sense of diffused light.
- **Drizzle and Light Precipitation:** Stratus clouds are known for producing light drizzle or mist. The water droplets within the cloud layer are too small to fall as heavier rain, but they create a persistent light precipitation that can dampen the landscape.

A Sign of Gloom or a Passing Phase?

While often associated with dreary weather, stratus clouds can offer clues about atmospheric conditions and potential changes:

- **Persistent Drizzle:** A persistent layer of stratus clouds accompanied by steady drizzle can indicate stable, cool air at low levels. This weather pattern can linger for extended periods, particularly during the winter months.
- **Breaking Stratus:** The gradual thinning and breaking up of a stratus cloud layer can signify improving weather conditions. As the air warms or dries, the cloud layer becomes less stable and eventually dissipates, potentially leading to clearer skies.
- **Preceding Other Cloud Types:** The presence of stratus clouds can sometimes precede the arrival of other cloud formations associated with more significant weather systems. For instance, a thickening stratus layer might indicate the approach of a warm front, which could bring clouds and precipitation.

Beyond Gloom: The Broader Role of Stratus Clouds

Despite their association with gloomy weather, stratus clouds play a vital role in Earth's climate system:

- **Greenhouse Effect:** Stratus clouds, composed entirely of water droplets, can trap heat radiating from Earth's surface, contributing to a warming effect. However, the net effect depends on the thickness and duration of the cloud cover. Thicker stratus layers can have a more significant warming effect by trapping more heat, while thinner layers might allow some heat to escape.
- **Albedo Effect:** Stratus clouds also influence the amount of solar radiation reaching Earth's surface. A thick stratus layer can reflect a significant portion of incoming sunlight back towards space, contributing to a cooling effect. Understanding the interplay between these opposing effects is crucial for assessing the overall impact of stratus clouds on climate.

Enhancing Your Stratus Spotting Skills

Here are some tips to become a more adept at spotting stratus clouds

- **Observe Cloud Thickness:** Pay attention to the thickness of the stratus layer. Thin stratus sheets might allow faint outlines of the sun or moon to be visible, while thicker layers will completely obscure them. The thickness of the cloud layer can also indicate the potential for precipitation – thicker layers are more likely to produce persistent drizzle.
- **Monitor Cloud Changes:** Track the changes in the stratus cloud cover over time. A persistent stratus layer suggests stable atmospheric conditions, while a gradual thinning or breaking up of the layer could signal improving weather.
- **Identify Accompanying Weather:** Observe the presence of other weather elements like fog or mist. Stratus clouds often coexist with fog, particularly in the morning hours, as fog can lift and transform into a stratus layer.

By observing and interpreting the characteristics of stratus clouds, you gain valuable insights into the processes unfolding near Earth's surface. Their presence can indicate radiative cooling, air mass movements, and even offer clues about the stability of the lower atmosphere.

NIMBOSTRATUS

When you look skyward on a dreary day, a seemingly endless sheet of dark gray clouds might dominate the scene, obscuring the sun and casting a veil of persistent rain. This uniform blanket of moisture signifies the presence of nimbostratus clouds, the low-level rain producers in the cloud classification system. Often associated with steady precipitation and gloomy weather, nimbostratus clouds play a crucial role in Earth's hydrological cycle. This section delves into the world of nimbostratus, exploring their formation, characteristics, and the stories they reveal about the atmospheric processes at play.

The Realm of Nimbostratus

Nimbostratus clouds reside in the lower troposphere, occupying a space between the low-level stratus clouds and the mid-altitude altostratus and altocumulus clouds. They typically base themselves between 0.6 and 3 kilometers (0.4-1.8 miles) above Earth's surface, encountering temperatures ranging from -5°C to 15°C (23°F to 59°F). Due to these moderate temperatures, nimbostratus clouds are composed primarily of water droplets, with the potential for ice crystals to form in the colder upper portions of the cloud layer. In some instances, particularly during winter months, the base of a nimbostratus cloud can brush the ground, leading to a transition into fog or mist.

Formation

The formation of nimbostratus clouds hinges on large-scale lifting and saturation of moist air within the lower troposphere. This lifting can occur due to several mechanisms:

- **Warm Fronts:** As warm air masses ascend over cooler air masses along a warm front, the rising warm air cools adiabatically (cools due to decreasing pressure) and eventually reaches saturation. This widespread condensation leads to the formation of a vast nimbostratus cloud layer, often accompanied by continuous precipitation.
- **Large-Scale Convergence:** The convergence of air masses at low levels can trigger lifting and adiabatic cooling, leading to saturation and the formation of nimbostratus clouds. This convergence can be associated with weather systems like cyclones or areas of low pressure.
- **Orographic Lifting:** As moist air is forced to rise over mountains (orographic lifting), it cools adiabatically and condenses, forming nimbostratus clouds on the windward side of the mountains. This process can lead to persistent precipitation on the slopes facing the wind.

Appearance

Nimbostratus clouds are known for their featureless, uniform appearance, often resembling a vast, dark gray sheet that obscures the sky. Here's a closer look at their key characteristics:

- **Sheet-Like Structure:** Nimbostratus clouds appear as a featureless, grayish-black sheet that can cover the entire sky. They effectively block sunlight, casting a gloomy and monotonous light over the landscape.
- **Sun and Moon Visibility:** Nimbostratus clouds are thick enough to completely obscure the sun or moon, leaving the sky a featureless expanse of gray.
- **Continuous Precipitation:** Nimbostratus clouds are renowned for producing continuous and often heavy precipitation, primarily in the form of rain or drizzle. The water droplets within the cloud

layer are large enough to fall as sustained precipitation, unlike the light drizzle associated with stratus clouds. In colder temperatures, nimbostratus clouds can also produce snow or sleet.

- **Virga:** Occasionally, precipitation from nimbostratus clouds might evaporate before reaching the ground, appearing as wispy trails known as virga. However, virga is less common with nimbostratus clouds compared to altostratus clouds due to the larger droplet size and higher likelihood of reaching the surface.

A Sign of Persistent Rain or Something More?

While primarily associated with persistent rain, nimbostratus clouds can offer clues about the atmospheric conditions and potential weather developments:

- **Steadiness of Precipitation:** The consistency of precipitation from nimbostratus clouds can provide insights into the stability of the weather system. Steady, continuous rain often suggests a persistent warm front or a well-developed area of low pressure.
- **Cloud Thickness:** The thickness of the nimbostratus layer can indicate the potential duration of the precipitation. Thicker layers are likely to produce longer periods of rain compared to thinner layers.
- **Clearing Skies:** The gradual thinning and breaking up of a nimbostratus cloud layer can signal the passage of a warm front and a transition to drier weather conditions.

Beyond Gloom: The Broader Role of Nimbostratus Clouds

Despite their association with gloomy weather, nimbostratus clouds play a vital role in Earth's climate. Here are a few of them

- **Hydrological Cycle:** Nimbostratus clouds are essential contributors to Earth's hydrological cycle. The continuous precipitation they produce replenishes freshwater resources, nourishes vegetation, and contributes to river flow and groundwater recharge.
- **Radiative Effects:** Nimbostratus clouds, composed primarily of water droplets, can trap heat radiating from Earth's surface,

contributing to a warming effect. However, the net effect depends on the thickness and duration of the cloud cover. Thicker nimbostratus layers can have a more significant warming effect by trapping more heat, while thinner layers might allow some heat to escape back towards space. Additionally, the thick cloud cover associated with nimbostratus clouds can also block incoming solar radiation, contributing to a cooling effect. Understanding the interplay between these opposing effects is crucial for assessing the overall impact of nimbostratus clouds on climate.

Enhancing Your Nimbostratus Spotting Skills

Here are some tips to become a more adept nimbostratus cloud spotter:

- **Observe Cloud Thickness and Base:** Pay close attention to the thickness and base of the nimbostratus layer. Thicker layers will completely obscure the sun and moon, while thinner layers might allow faint outlines to be visible. A very low cloud base, sometimes brushing the ground, can indicate the possibility of encountering fog or mist alongside the nimbostratus cloud cover.
- **Monitor Precipitation:** The type and intensity of precipitation can offer clues about the nimbostratus cloud layer. Continuous rain or drizzle is a hallmark of nimbostratus, while snow or sleet might be present in colder temperatures. Observe the consistency of the precipitation – a steady and unwavering downpour suggests a persistent nimbostratus layer.
- **Look for Signs of Change:** Track the changes in the nimbostratus cloud cover over time. A gradual thinning and breaking up of the layer can signal the passage of a warm front and potentially drier weather on the horizon.

By observing and interpreting the characteristics of nimbostratus clouds, you gain valuable insights into the large-scale processes unfolding within the lower troposphere. Their presence can indicate the movement of warm fronts, areas of low pressure, and orographic lifting, all of which influence weather patterns.

STRATOCUMULUS

When you look skyward on a day that seems undecided between clear and cloudy, you might encounter a peculiar sight – a patchy canvas dotted with rounded clumps of white or gray clouds. This ever-shifting formation signifies the presence of stratocumulus, the cloud type known for its lumpy texture and variable appearance. Often associated with fair weather conditions, stratocumulus clouds offer valuable insights into atmospheric stability and can even hint at potential changes on the horizon. This section delves into the world of stratocumulus, exploring their formation, characteristics, and the stories they reveal about the conditions above.

The Realm of Stratocumulus

Stratocumulus clouds reside in the lower troposphere, sharing airspace with stratus clouds but typically inhabiting slightly higher altitudes. They generally range from 0.5 to 3 kilometers (0.3-1.8 miles) above Earth's surface, encountering temperatures between -5°C to 15°C (23°F to 59°F). Due to these moderate temperatures, stratocumulus clouds can be composed of a mixture of water droplets and ice crystals, depending on the specific atmospheric conditions.

Formation

The formation of stratocumulus clouds hinges on the lifting and cooling of moist air within the lower troposphere. Here's a closer look at the key mechanisms involved:

- **Adiabatic Cooling:** As air ascends due to processes like convection or forced lifting over mountains (orographic lifting), it cools adiabatically (cools due to decreasing pressure). When this rising air reaches a level where its temperature dips below the dew point temperature, condensation occurs, leading to the formation of cloud droplets or ice crystals. Unlike widespread lifting that produces uniform stratus clouds, the lifting associated with stratocumulus can be more localized, resulting in a patchy distribution of cloud formations.
- **Convergence Zones:** The horizontal convergence of air masses at low levels can trigger lifting and adiabatic cooling, creating conditions favorable for stratocumulus formation. Convergence zones can be associated with weather systems like sea breezes or areas of low pressure.
- **Instability within Stable Air:** Occasionally, even within a stable atmosphere, weak updrafts and fluctuations in temperature can trigger localized condensation and the formation of patchy stratocumulus clouds.

Appearance

Stratocumulus clouds are renowned for their variable and often picturesque appearance. Here's a closer look at their key characteristics:

- **Patchy Distribution:** Stratocumulus clouds appear as a collection of individual rounded clumps or rolls, often arranged in a patchy or wave-like pattern. The distribution of these cloud elements can be sparse or dense, creating a constantly changing celestial tapestry.
- **Varied Texture:** The texture of stratocumulus clouds can range from smooth and billowy to lumpy and uneven. This variability depends on the strength of updrafts and the presence of turbulence within the layer.

- **Sun and Moon Visibility:** Stratocumulus clouds can vary in their ability to obscure the sun or moon. Thin patches might allow the sun to shine through with a slightly diffused light, while thicker patches or dense collections of stratocumulus can create a dappled sunlight effect or even partially obscure the sun or moon.
- **Variable Precipitation:** Stratocumulus clouds are not typically associated with significant precipitation. However, in some cases, particularly with thicker patches, light drizzle or virga (wispy trails of evaporating precipitation) might occur.

Fair Weather or a Sign of Change?

Stratocumulus clouds can be associated with both fair weather and approaching weather systems:

- **Fair Weather:** Patchy stratocumulus with clear spaces between the cloud elements often signifies stable atmospheric conditions. This is a common sight on days transitioning from clear mornings to partly cloudy afternoons.
- **Marine Layer:** Over oceans, extensive layers of stratocumulus can form due to cool air masses moving over warmer water. These marine stratocumulus layers can persist for extended periods and are often associated with fog formation near the coast.
- **Preceding Other Cloud Types:** In some instances, the presence of stratocumulus can precede the arrival of other cloud formations associated with more significant weather systems. For instance, a thickening layer of stratocumulus with a more uniform appearance could indicate the approach of a warm front.

The Broader Role of Stratocumulus Clouds

While their patchy appearance might not suggest a grand role, stratocumulus clouds play a vital role in Earth's climate system:

- **Albedo Effect:** Stratocumulus clouds, particularly those with a white and reflective appearance, can reflect a significant portion of incoming solar radiation back towards space, contributing to a cooling effect. This effect is particularly important over vast ocean regions where stratocumulus clouds can form extensive layers.

Understanding the reflective properties of stratocumulus clouds is crucial for assessing their overall impact on climate.

- **Suppression of Convection:** In some situations, stratocumulus cloud layers can act as a lid, trapping cooler air near the surface and suppressing the development of convection. This can have implications for regional weather patterns and can influence the formation of other cloud types.
- **Marine Ecosystems:** Extensive marine stratocumulus layers can play a role in regulating ocean temperatures by reflecting sunlight back towards space. This can indirectly influence marine ecosystems by affecting the distribution of heat and light in the upper ocean layers.

Enhancing Your Stratocumulus Spotting Skills

Here are some tips to become a more adept stratocumulus cloud spotter:

- **Observe Cloud Distribution and Texture:** Pay close attention to the arrangement and texture of the stratocumulus cloud elements. Sparse, rounded clumps with smooth textures often suggest stable conditions, while denser patches with a more lumpy or billowy appearance might indicate the presence of weak turbulence.
- **Monitor Changes Over Time:** Track the evolution of the stratocumulus cloud layer. Are the patches thickening or thinning? Are they becoming more or less extensive? Observing these changes can offer clues about potential shifts in atmospheric conditions.
- **Identify Accompanying Weather:** Look for the presence of other weather elements that might provide context. For instance, the presence of stratocumulus alongside fog formation near a coastline suggests a marine layer.

By observing and interpreting the characteristics of stratocumulus clouds, you gain valuable insights into the subtle processes unfolding within the lower troposphere. Their presence can indicate lifting mechanisms, areas of convergence, and even hint at the stability of the atmosphere.

CHAPTER 7

VERTICAL CLOUDS

CUMULUS

When you look skyward on a bright summer day, you might encounter a scene straight out of a classic painting – a vibrant blue canvas dotted with fluffy white shapes resembling cotton balls. These delightful formations are cumulus clouds, the quintessential fair-weather friends in the cloud classification system. Often associated with sunny skies and pleasant conditions, cumulus clouds offer a glimpse into atmospheric convection and can even provide clues about potential weather developments. This section delves into the world of cumulus, exploring their formation, characteristics, and the stories they reveal about the conditions aloft.

The Home of Cumulus

Cumulus clouds reside in the lower troposphere, typically inhabiting altitudes between 0.5 and 2 kilometers (0.3-1.2 miles) above Earth's surface. They encounter temperatures ranging from 5°C to 20°C (41°F to 68°F). Due to these mild temperatures, cumulus clouds are composed entirely of water droplets. In rare instances, at the very tops of the tallest cumulus clouds, ice crystals might begin to form as temperatures dip below freezing.

Formation

The formation of cumus clouds hinges on a fundamental atmospheric process – convection. As the sun heats Earth's surface, pockets of air near the ground become warmer and less dense. Buoyed by their buoyancy, these warm air parcels begin to rise through the cooler surrounding air. As they ascend, they experience adiabatic cooling (cooling due to decreasing pressure). When the rising air reaches a level where its temperature falls below the dew point, condensation occurs, leading to the formation of tiny water droplets. These water droplets coalesce to form the visible cloud – a cumulus cloud.

Appearance

Cumulus clouds are renowned for their distinctive and often delightful appearance. Here's a closer look at their key characteristics:

- **Puffy Texture:** Cumulus clouds appear as individual, puffy masses with rounded tops and flat bases. Their resemblance to cotton balls or pillows has earned them the nickname "fair-weather clouds."
- **Variable Size:** Cumulus clouds can vary considerably in size. Small cumulus humilis clouds might be just a few hundred meters across, while towering cumulus congestus clouds can reach impressive heights, exceeding several kilometers.
- **Sharp Outlines:** Cumulus clouds typically have well-defined, sharp outlines, particularly on their sunny sides. This is because condensation within the cloud occurs primarily along the rising edges of the air parcels.
- **Fair-Weather Association:** Cumulus clouds are most commonly associated with fair weather conditions. Their presence often indicates stable atmospheric conditions with localized areas of rising air.

Fair Weather or a Sign of More?

While primarily associated with fair weather, cumulus clouds can offer clues about atmospheric conditions and potential changes:

- **Cloud Development:** The stage of development of a cumulus cloud can provide insights into atmospheric stability. Small, puffy cumulus humilis clouds suggest stable conditions, while the towering cumulus congestus clouds with anvil-shaped tops can indicate stronger updrafts and the potential for the development of cumulonimbus clouds, which bring showers and thunderstorms.
- **Cloud Dispersal:** The gradual dissipation of cumulus clouds throughout the day can signify a weakening of convection and a return to clear skies.
- **Cloud Organization:** The arrangement of cumulus clouds can sometimes offer clues about wind direction at higher altitudes. Long lines or streaks of cumulus clouds aligned in a particular direction might indicate the presence of strong winds aloft.

The Broader Role of Cumulus Clouds

Despite their association with fair weather, cumulus clouds play a vital role in Earth's climate system:

- **Reflecting Sunlight:** Cumulus clouds, particularly the larger and more voluminous cumulus congestus clouds, can reflect a portion of incoming solar radiation back towards space, contributing to a cooling effect. However, the net effect depends on the size, number, and duration of the clouds.
- **Shaping Weather Patterns:** The presence and development of cumulus clouds can influence the formation of more complex weather systems. For instance, the towering cumulus congestus clouds can transition into cumulonimbus clouds, which are associated with thunderstorms and precipitation.
- **Seeding Precipitation:** Cumulus clouds, particularly the larger varieties, can act as seed clouds for precipitation. Tiny ice crystals that form at the tops of these clouds can initiate the Bergeron-Findeisen process, leading to the growth of larger ice particles and eventual precipitation.

Enhancing Your Cumulus Spotting Skills

Here are some tips to become a more adept cumulus cloud spotter:

- **Observe Cloud Size and Development:** Pay close attention to the size and stage of development of cumulus clouds. Small, puffy cumulus humilis suggest stable conditions, while towering cumulus congestus with anvil-shaped tops indicate stronger updrafts. Look for signs of growth or dissipation throughout the day to understand the evolving atmospheric conditions.
- **Identify Cloud Organization:** Observe if the cumulus clouds are scattered randomly or arranged in lines or streaks. Linear arrangements might hint at the presence of strong winds aloft, influencing the movement and organization of the rising air parcels.
- **Monitor Cloud Shadows:** The shadows cast by cumulus clouds on the ground can reveal their relative height and three-dimensional structure. Taller cumulus congestus will cast more prominent and well-defined shadows compared to smaller cumulus humilis clouds.

By observing and interpreting the characteristics of cumulus clouds, you gain valuable insights into the workings of atmospheric convection. Their presence signifies localized areas of rising air, and their development can reveal the strength and stability of these updrafts.

CUMULONIMBUS

When you look skyward on a restless afternoon, a dramatic transformation might unfold before your eyes. Fluffy cumulus clouds begin to merge and darken, their cotton-ball forms morphing into towering giants with an ominous presence. These ever-evolving titans of the sky are cumulonimbus clouds, the harbingers of storms and the most dramatic cloud formation in the classification system. Cumulonimbus clouds are the powerhouses of the atmosphere, capable of producing a variety of weather phenomena, from heavy rain and hail to lightning and even tornadoes. This section delves into the world of cumulonimbus, exploring their formation, characteristics, and the stories they reveal about the potent atmospheric processes at play.

The Vertical Realm of Cumulonimbus

Cumulonimbus clouds are the undisputed giants of the lower atmosphere. They reside in the vertical realm, with their bases nestled within the lower troposphere, often starting around 1 to 2 kilometers (0.6-1.2 miles) above Earth's surface. However, their towering tops can pierce through the troposphere, reaching heights of 10 to 20 kilometers (6.2-12.4 miles) in extreme cases. Due to the vast range of altitudes they occupy, cumulonimbus clouds encounter a wide temperature spectrum, ranging from mild temperatures near the base to frigid conditions at the top, where ice crystals dominate.

Formation

The formation of cumulonimbus clouds hinges on a powerful atmospheric phenomenon – vigorous convection. As strong sunlight heats the Earth's surface, pockets of air near the ground become significantly warmer and less dense. These buoyant air parcels rise rapidly through the cooler surrounding air, experiencing adiabatic cooling (cooling due to decreasing pressure) as they ascend. The rising air cools, eventually reaching a level where its temperature dips below the dew point, triggering condensation. However, unlike the localized updrafts associated with fair-weather cumulus clouds, the updrafts within cumulonimbus clouds are much stronger and more sustained, leading to the formation of a massive, vertically oriented cloud mass.

As the rising air continues its ascent within the cumulonimbus cloud, the water droplets further cool and condense, eventually transitioning into ice crystals as temperatures plummet. The interplay between water droplets and ice crystals within the cloud plays a crucial role in the development of precipitation and electrical charges, ultimately leading to the various forms of stormy weather associated with cumulonimbus clouds.

Appearance

Cumulonimbus clouds are renowned for their dramatic and imposing appearance. Here's a closer look at their key characteristics:

- **Towering Structure:** Cumulonimbus clouds appear as massive, vertically developed cloud formations with a towering, anvil-shaped top that can stretch for several kilometers.
- **Dark and Ominous:** The upper portions of cumulonimbus clouds are typically dark and ominous due to the presence of water droplets and ice crystals that effectively block sunlight.
- **Flat Base:** The base of a cumulonimbus cloud often appears flat and may exhibit ragged edges or precipitation streaks. In some instances, the base can be so low that it brushes the ground, leading to the formation of virga (wispy trails of evaporating precipitation) or even fog.
- **Rapidly Evolving:** Cumulonimbus clouds are not static formations. They undergo rapid changes in size, shape, and intensity throughout their lifecycle, reflecting the dynamic nature of the atmospheric processes at play.

The Story Cumulonimbus Clouds Tell

The presence of cumulonimbus clouds signifies a significant shift in atmospheric conditions and foreshadows the arrival of stormy weather:

- **Heavy Precipitation:** Cumulonimbus clouds are the primary producers of heavy rain, hail, and thunderstorms. The vigorous updrafts within these clouds allow for the sustained growth of water droplets and ice crystals, which eventually become too heavy to remain suspended and fall as precipitation.

- **Lightning and Thunder:** The interaction between water droplets and ice crystals within cumulonimbus clouds leads to the separation of electrical charges, ultimately resulting in the powerful display of lightning and the accompanying sound of thunder.
- **Strong Winds:** Cumulonimbus clouds can be associated with strong downdrafts and turbulence, particularly in the vicinity of heavy precipitation.

The Broader Role of Cumulonimbus Clouds

While primarily associated with storms, cumulonimbus clouds play a vital role in Earth's climate system:

- **Vertical Redistribution of Heat and Moisture:** Cumulonimbus clouds act as giant conveyor belts, transporting heat and moisture from the Earth's surface to higher altitudes in the troposphere. This vertical redistribution contributes to maintaining the overall balance of energy within the atmosphere. The transport of moisture by cumulonimbus clouds is particularly crucial for replenishing freshwater resources through precipitation events.

- **Shaping Weather Systems:** Cumulonimbus clouds can play a significant role in the development and evolution of larger weather systems. Their formation and associated precipitation patterns can influence frontal boundaries and contribute to the overall circulation patterns within the atmosphere.
- **Cleansing the Atmosphere:** Cumulonimbus clouds, through the process of precipitation, can help cleanse the atmosphere of dust, pollen, and other airborne particles. This scavenging effect contributes to maintaining air quality.

Enhancing Your Cumulonimbus Spotting Skills

Given their dramatic appearance and association with stormy weather, cumulonimbus clouds are some of the easiest clouds to identify in the sky. However, here are some tips to become a more adept cumulonimbus cloud spotter:

- **Observe Cloud Development:** Pay close attention to the vertical development of the cloud. Cumulonimbus clouds exhibit a distinct towering structure with a well-defined anvil-shaped top, unlike the flatter appearance of other cloud types.
- **Monitor for Rapid Changes:** Cumulonimbus clouds are constantly evolving. Look for signs of rapid growth, darkening of the upper portions, and the development of an anvil shape. These changes often indicate an intensifying storm.
- **Identify Associated Weather:** If you spot a cumulonimbus cloud, be aware of the potential for heavy rain, hail, lightning, and strong winds. Monitor weather reports or listen for thunder to stay informed about the storm's development.

By observing and interpreting the characteristics of cumulonimbus clouds, you gain valuable insights into the powerful forces at play within the atmosphere. Their presence signifies vigorous convection, the potential for heavy precipitation, and the possibility of electrical activity.

PART 3

More Insights and Cloud Spotting Tips

CHAPTER 8

CLOUD COMPOSITIONS AND MORE

ICE CRYSTAL CLOUDS

When you look skyward on a crisp winter morning, the vast expanse of blue might be adorned with a dazzling display unlike any other. Delicate wisps, shimmering veils, or iridescent patches of color grace the high reaches of the atmosphere, forming a tapestry of crystalline beauty. These are the ice crystal clouds, residing in the uppermost reaches of the troposphere and beyond, where temperatures plummet and water vapor transforms into a frozen spectacle. This section delves into the world of ice crystal clouds, exploring their formation, characteristics, and the stories they reveal about the frigid conditions high above our planet.

Ice crystal clouds, unlike their lower-altitude counterparts composed primarily of water droplets, exist in the frigid realm of the upper troposphere and stratosphere. These lofty domains begin at around 5 to 13 kilometers (3-8 miles) above Earth's surface, with some cloud formations extending even higher, reaching the mesosphere at altitudes exceeding 50 kilometers (31 miles). Due to these extreme temperatures, ranging from -40°C to -85°C (-40°F to -121°F), water vapor in the atmosphere readily sublimates (transitions directly from a gas to a solid) into tiny ice crystals, forming the building blocks of these high-altitude wonders.

Formation

The formation of ice crystal clouds hinges on various atmospheric mechanisms, depending on the specific cloud type. Here's a look at some key processes involved:

- **Sublimation:** As mentioned earlier, the primary driver for ice crystal cloud formation is the sublimation of water vapor at high altitudes. When air masses ascend and encounter the frigid temperatures of the upper troposphere, water vapor bypasses the liquid phase entirely and transforms directly into ice crystals.

- **Lifting and Adiabatic Cooling:** Similar to the formation of lower-altitude cloud types, lifting and adiabatic cooling (cooling due to decreasing pressure) can play a role in ice crystal cloud formation. Rising air parcels cool, eventually reaching a point where saturation occurs with respect to ice, leading to the formation of ice crystals. This mechanism is particularly important for cirrocumulus and cirrostratus clouds.
- **In-Cloud Processes:** Within existing ice crystal clouds, additional ice crystals can form through the Bergeron-Findeisen process. This process involves the coexistence of supercooled water droplets (water droplets remaining liquid at temperatures below freezing) and ice crystals within the cloud. The ice crystals, due to their efficient vapor pressure, grow at the expense of the supercooled droplets, leading to further ice crystal formation.

Appearance

Ice crystal clouds are renowned for their ethereal beauty and diverse appearances. Here's a closer look at their key characteristics:

- **Wispy and Delicate:** Ice crystal clouds often appear as wispy streaks, feathery plumes, or thin veils across the sky. Their delicate structure reflects the sparse distribution of ice crystals within these formations.
- **Halo Phenomena:** Ice crystals in cirrus clouds can refract and diffract sunlight, creating a dazzling array of halo phenomena. These include halos (circular rings around the sun or moon), sundogs (bright spots appearing beside the sun), and light pillars (vertical shafts of light extending from the sun or moon).
- **Iridescent Colors:** Under certain conditions, ice crystal clouds can exhibit a mesmerizing display of iridescent colors. These colors arise due to the diffraction of sunlight by the ice crystals, creating a spectrum of pastel hues that can transform a seemingly ordinary cloud into a breathtaking spectacle.
- **Variable Visibility:** The visibility of ice crystal clouds can vary depending on the density and thickness of the cloud layer. Thin cirrus clouds might be nearly transparent, allowing the sun to shine through, while thicker cirrostratus clouds can obscure the sun and create a hazy appearance in the sky.

The Role of Ice Crystal Clouds in Earth's Climate

While often admired for their ethereal beauty, ice crystal clouds play a crucial role in Earth's climate system:

- **Radiative Effects:** Ice crystal clouds can have a complex impact on Earth's radiation balance. Thin cirrus clouds, particularly at higher altitudes, can act as heat traps, allowing incoming solar radiation to pass through but trapping some of the outgoing infrared radiation emitted by Earth's surface. Thicker cirrostratus clouds, on the other hand, can reflect a significant portion of incoming solar radiation back towards space, contributing to a cooling effect. Understanding the net radiative impact of ice crystal clouds is an ongoing area of research in climate science.
- **Seeding Precipitation:** Ice crystals in cirrostratus clouds can play a role in initiating precipitation in mixed-phase clouds (clouds containing both ice crystals and water droplets) at lower altitudes. The Bergeron-Findeisen process, mentioned earlier, can lead to the growth of larger ice crystals within cirrostratus clouds. These larger ice crystals can then fall through the atmosphere and collide with supercooled water droplets in lower-level clouds. The collision triggers the freezing of the supercooled droplets and initiates the growth of even larger precipitation particles, eventually leading to rain or snow.
- **Influencing Atmospheric Circulation:** The presence and properties of ice crystal clouds can influence atmospheric circulation patterns. High-altitude cirrus clouds can trap heat near the tropopause (the boundary between the troposphere and stratosphere), potentially affecting the jet stream and storm development.

MIXED-PHASE CLOUDS

As you peer skyward on a day with a dynamic atmosphere, you might encounter a cloud formation that defies easy categorization. It possesses characteristics of both the fluffy cumulus clouds and the wispy cirrus clouds – a seemingly contradictory blend. These are mixed-phase clouds, a fascinating category where water droplets and ice crystals coexist in a delicate balance. Their presence signifies a specific set of atmospheric conditions and plays a crucial role in shaping precipitation patterns and influencing Earth's climate. This section delves into the world of mixed-phase clouds, exploring their formation, characteristics, and the stories they reveal about the interplay between warmth and cold within the atmosphere.

Mixed-phase clouds reside in a transitional zone within the troposphere, typically between 0°C to -40°C (32°F to -40°F). This temperature range allows for the coexistence of water droplets, which remain liquid in a supercooled state below freezing, and ice crystals, which readily form at these temperatures. The specific altitude of mixed-phase clouds can vary depending on factors like temperature lapse rate (change in temperature with altitude) and moisture content of the rising air parcel.

Formation

The formation of mixed-phase clouds hinges on a complex interplay between lifting mechanisms, cooling processes, and the competition between water droplets and ice crystals for atmospheric moisture. Here's a breakdown of the key factors involved:

- **Lifting and Adiabatic Cooling:** Similar to the formation of other cloud types, lifting of air parcels plays a crucial role. As air rises, it cools adiabatically, eventually reaching a saturation point where condensation occurs. Depending on the temperature at the saturation level, this condensation can result in the formation of either water droplets (if the temperature is above freezing) or ice crystals (if the temperature is below freezing).
- **Supercooled Water Droplets:** In some instances, particularly within warmer mixed-phase clouds, condensation can lead to the formation of supercooled water droplets. These droplets remain liquid despite being at subfreezing temperatures due to the lack of ice nucleation sites (particles around which ice crystals can form).
- **The Bergeron-Findeisen Process:** A critical process within mixed-phase clouds is the Bergeron-Findeisen process. This process hinges on the differing vapor pressures of water droplets and ice crystals. Due to their lower vapor pressure, ice crystals are more efficient at attracting water vapor compared to supercooled droplets. As a result, ice crystals grow at the expense of

supercooled droplets, leading to the further growth of ice crystals and a gradual depletion of liquid water within the cloud.

Appearance

Mixed-phase clouds exhibit a diverse range of appearances, reflecting the dynamic interplay between water droplets and ice crystals within them. Here's a closer look at their key characteristics:

- **Variable Texture:** The texture of mixed-phase clouds can vary depending on the relative abundance of water droplets and ice crystals. Clouds dominated by water droplets might appear puffier and more similar to cumulus clouds, while those with a higher ice crystal content might exhibit wispy streaks or a more cirrus-like appearance.
- **Shifting Forms:** Mixed-phase clouds are not static formations. The competition between water droplets and ice crystals, along with changes in temperature and updraft strength, can lead to continuous changes in their shape and texture.
- **Precipitation Potential:** The presence of both water droplets and ice crystals within mixed-phase clouds signifies the potential for precipitation development. The Bergeron-Findeisen process can lead to the growth of large ice particles, which can then fall as snow or initiate rain formation through collision-coalescence with supercooled droplets.

HOW HUMAN ACTIVITY AFFECTS CLOUD FORMATION

Gazing at the ever-changing canvas of clouds, it's easy to overlook the subtle influence humans exert on these formations. Our activities, particularly the burning of fossil fuels and agricultural practices, are altering the delicate balance within the atmosphere, impacting cloud formation and behavior in various ways.

Understanding Cloud Formation

Before exploring human influences, let's revisit the natural processes governing cloud formation. Clouds form when invisible water vapor condenses into visible water droplets or ice crystals. This condensation hinges on two key factors:

- **Temperature:** Rising air parcels cool due to decreasing pressure. When the temperature reaches a critical point (dew point or frost point), condensation occurs, marking the birth of a cloud.
- **Aerosol Concentration:** Tiny airborne particles called aerosols act as microscopic platforms for water vapor condensation. These "condensation nuclei" come from natural sources like dust, sea spray, and volcanic ash. The abundance and type of aerosols influence the size, number, and properties of cloud droplets.

Human Activities and Aerosol Emissions

Human activities significantly alter the concentration and composition of aerosols in the atmosphere. Here's a closer look at some key contributors:

- **Fossil Fuel Combustion:** Burning coal, oil, and natural gas releases pollutants, including aerosols composed of sulfates, nitrates, and black carbon. These anthropogenic aerosols can act as efficient condensation nuclei, potentially increasing the number of cloud droplets but often making them smaller.
- **Agricultural Practices:** Tilling, animal husbandry, and fertilizers contribute to dust particles and ammonia emissions, both of which influence aerosol concentration and indirectly lead to the formation of certain aerosols.
- **Biomass Burning:** Forest fires and burning agricultural residue release aerosols, impacting cloud formation patterns in affected regions.

How Human Activities Affect Clouds

The increase in aerosols due to human activities has a cascading effect on clouds, influencing their properties and behavior:

- **Altered Cloud Droplet Size:** Increased aerosol concentration can lead to more numerous but smaller cloud droplets. This can affect the overall reflectivity of clouds, potentially impacting how much solar radiation they reflect back towards space.

- **Indirect Effects on Precipitation:** Smaller cloud droplets tend to take longer to coalesce and fall as rain. This can delay the onset of precipitation and potentially alter storm formation patterns.
- **Lifetime and Dissipation:** Changes in cloud droplet size and composition can influence how long clouds persist in the atmosphere. Smaller droplets might evaporate more readily, leading to shorter cloud lifetimes and affecting regional weather patterns.

- **Cloud Type Specificity:** The influence of human-induced aerosols can vary depending on the specific cloud type. High-altitude cirrus clouds might be less affected compared to lower-level cumulus clouds.
- **Regional Variations:** The impact can exhibit significant regional variations. Areas with heavy industry or extensive agriculture might experience more pronounced effects compared to pristine environments.
- **Scientific Uncertainty:** Understanding the complex interplay between human activities, aerosols, and clouds remains an active area of research. Quantifying the precise effects and their long-term implications for weather patterns and climate is an ongoing challenge.

As we move forward, understanding and mitigating the human influence on clouds is crucial for addressing climate change. Here are some ongoing efforts:

- **Aerosol Research:** Scientists are actively researching the impact of different types of aerosols on cloud formation and behavior. This research is crucial for developing strategies to reduce the emissions of harmful aerosols.
- **Climate Modeling:** Climate models are being refined to better account for the complex interactions between aerosols, clouds, and climate. This will improve our ability to predict the future impacts of human activities on cloud formation and weather patterns.

By understanding the invisible hand humans exert on clouds, we gain a deeper appreciation for the delicate balance within the atmosphere.

As we strive for a sustainable future, mitigating our impact on cloud formation is essential for ensuring the health of our planet and its ever-changing weather patterns.

CHAPTER 9

CLOUD PHOTOGRAPHY

CHOOSING THE RIGHT CAMERA SETTINGS FOR CLOUD PHOTOGRAPHY

As a cloud enthusiast, you've likely experienced the urge to translate the breathtaking beauty of the skyscape into lasting images. Cloud photography presents a unique challenge – capturing the ever-changing forms and delicate details of these celestial formations. This section equips you with the knowledge to choose the ideal camera settings, allowing you to transform fleeting moments into captivating photographs.

Understanding the Core Settings

Before venturing out with your camera, it's crucial to grasp the fundamental trio of camera settings that govern exposure and image quality:

- **Aperture:** Represented by f-stops (e.g., f/2.8, f/8, f/16), aperture refers to the opening of the lens that controls the amount of light entering the camera. A wider aperture (lower f-stop number) allows more light in, creating a shallow depth of field (blurred background). Conversely, a narrower aperture (higher f-stop number) lets in less light but produces a deeper depth of field (everything in focus).
- **Shutter Speed:** Measured in seconds or fractions of a second (e.g., 1/125s, 1/250s, 1s), shutter speed determines the duration the camera sensor is exposed to light. A faster shutter speed captures a fleeting moment and freezes motion. Conversely, a slower shutter speed allows more light to enter but can blur moving elements within the scene.
- **ISO:** Expressed in numerical values (e.g., ISO 100, ISO 400, ISO 1600), ISO controls the camera sensor's sensitivity to light. Higher ISO values increase sensitivity, allowing you to capture images in low-light conditions. However, higher ISO settings also introduce digital noise, which can affect image quality.

Capturing Cloud Details

When photographing clouds, the interplay between aperture and shutter speed becomes crucial:

- **Sharp Cloud Details:** For crisp, detailed cloud textures, prioritize a relatively narrow aperture (f/8 or higher). This ensures all parts of the cloud, from foreground to background, are in sharp focus.
- **Freezing Cloud Movement:** Clouds are not static formations. To capture sharp cloud details without blurring due to wind movement, consider using a faster shutter speed (around 1/250s or faster).
- **Creative Effects:** For a more dynamic feel, you can experiment with slower shutter speeds (around 1/125s or slower) to create a sense of movement in the clouds, especially when dealing with fast-moving formations or wispy streaks.
- **Balancing Light and Detail:** The ideal combination of aperture and shutter speed depends on the lighting conditions. In bright sunlight, you can use a narrower aperture and faster shutter speed to maintain detail and avoid overexposure. In low-light conditions, you might need to choose a wider aperture or a slower shutter speed (potentially introducing some blur) to capture a well-exposed image.

ISO and Cloud Photography

Finding the optimal ISO setting for cloud photography requires striking a balance between light sensitivity and image quality:

- **Prioritize Lower ISO:** Whenever possible, use the lowest ISO setting your camera allows (typically ISO 100 or 200). This ensures the cleanest image with minimal digital noise, crucial for capturing the subtle details and textures within the clouds.
- **Adjusting for Light:** In low-light conditions, such as sunrise or sunset, using a lower ISO might result in underexposed images. Gradually increase the ISO value while monitoring the image preview to ensure the clouds retain detail and the noise level remains acceptable.

- **Experiment for Artistic Effects:** While lower ISO is generally preferred, some photographers intentionally introduce noise for a more artistic or vintage feel. Experiment with higher ISO settings to see if the aesthetic aligns with your creative vision.

Additional Settings to Consider for Enhanced Cloud Photography

Beyond the core settings, consider these additional options for fine-tuning your cloud photographs:

- **Metering Mode:** Experiment with different metering modes, such as spot metering, to ensure proper exposure for the clouds, especially if they occupy a significant portion of the frame.
- **White Balance:** Depending on the lighting conditions, adjusting the white balance (e.g., daylight, cloudy) can ensure accurate color representation of the clouds, particularly at sunrise or sunset.
- **Use a Tripod:** A tripod provides stability and allows you to use slower shutter speeds without introducing camera shake, especially beneficial in low-light conditions.
- **Explore Bracketing:** Bracketing involves capturing multiple exposures of the same scene with slight variations in aperture, shutter speed, or ISO. This technique ensures you capture at least one well-exposed image, especially helpful in situations with challenging lighting.

Post-Processing Techniques

Once you've captured your breathtaking cloud formations, post-processing software can further enhance your images:

- **Basic Adjustments:** Adjustments like exposure, contrast, and highlights/shadows can help fine-tune the overall image balance and bring out details within the clouds.
- **Selective Adjustments:** Use masking tools to selectively adjust specific areas of the image, such as brightening the clouds while maintaining a natural look for the sky.
- **Color Enhancement:** Subtly adjust color saturation or use color grading tools to enhance the color palette of the clouds, particularly at sunrise or sunset.

- **Noise Reduction:** If you used a higher ISO setting, noise reduction software can help minimize the appearance of digital grain, especially in areas with flat color tones like the sky.
- **Sharpening:** Light sharpening can enhance the definition of cloud textures and details, particularly when combined with noise reduction techniques.

Remember, post-processing is a powerful tool, but use it subtly to preserve the natural beauty of the clouds. By understanding these camera settings, composition techniques, and post-processing tips, you'll be well-equipped to capture the breathtaking artistry of clouds. Embrace experimentation, explore different approaches, and let your creativity flow as you translate the ever-changing story of the skyscape into captivating photographs.

COMPOSITION TECHNIQUES FOR STUNNING CLOUD PHOTOS

Clouds, in their ever-changing dance across the sky, present a captivating subject for photographers. While mastering camera settings is crucial, translating the majesty of these formations into compelling images requires an eye for composition. This section delves into essential composition techniques that will elevate your cloud photography from snapshots to stunning works of art.

Rule of Thirds

Imagine dividing your camera's viewfinder or image preview into a 3x3 grid, creating nine equal squares. The rule of thirds suggests placing key elements of your composition, like dramatic cloud formations, along the intersecting lines or at the points where these lines meet. This approach creates a more balanced and visually appealing image compared to simply placing the clouds smack dab in the center.

Here's how to utilize the rule of thirds effectively:

- **Position the Main Cloud Formation:** Identify the most visually striking cloud feature you want to highlight. Use the rule of thirds grid to place it off-center, either to the left or right of the frame, along one of the intersecting lines.

- **Balance the Composition:** While the clouds are the main attraction, consider incorporating other elements to create visual balance. If placing a cloud formation off-center, use a captivating foreground element, like a lone tree or a winding road, positioned along an opposing intersecting line.
- **Experiment with Vertical Compositions:** The rule of thirds applies not just to horizontal but also to vertical compositions. When photographing towering cloud formations, consider placing the base of the clouds along the bottom intersecting line and the top along the top intersecting line.

Leading Lines

Leading lines are natural elements within a scene that draw the viewer's eye towards the main subject. In cloud photography, these lines can be anything from a winding road or a majestic mountain range in the foreground to the converging lines of fence posts or railroad tracks leading towards the horizon.

- **Utilize the Foreground:** Incorporate leading lines into the foreground of your image to create a sense of depth and guide the viewer's gaze towards the clouds in the background. Imagine the leading lines acting as arrows pointing towards the dramatic cloud formations you want to highlight.
- **Vary the Perspective:** Experiment with different camera angles to enhance the impact of leading lines. Get low to the ground to emphasize the foreground element and create a more dramatic perspective as the leading lines converge towards the clouds.
- **Be Mindful of Distractions:** While leading lines are powerful tools, ensure they don't lead the viewer's eye out of the frame entirely. Avoid placing distracting elements, like a telephone pole or a lamppost, directly along the leading lines.

The Power of the Foreground

While clouds are the stars of the show, incorporating a captivating foreground element can significantly enhance your image. Here's how a well-chosen foreground element can elevate your cloud photography:

- **Adding Context:** A foreground element can provide context and a sense of scale to your image. Imagine a lone tree silhouetted against a dramatic cloud formation, creating a sense of isolation and emphasizing the power of the sky.
- **Creating Depth:** Foreground elements add depth and dimension to an image. By including a foreground object with clear texture and detail, you create a layered composition that draws the viewer into the scene.
- **Juxtaposition:** Foreground elements can be used to create a sense of contrast with the clouds. For instance, a field of vibrant wildflowers in the foreground juxtaposed with dark, stormy clouds in the background can create a visually striking image.
- **Framing the Clouds:** Foreground elements can be used to frame the clouds, drawing attention to specific formations or creating a sense of mystery. Imagine a natural archway framing a portion of the cloud scape, highlighting its unique features.

Be Selective: While a foreground element can enhance your image, avoid cluttering the composition. Choose a single, well-defined foreground object that complements the clouds and doesn't compete for attention.

Negative Space

Negative space refers to the empty areas within a frame that are not occupied by the main subject. While it might seem counterintuitive, effectively using negative space can be a powerful tool in cloud photography.

- **Emphasizing the Clouds:** A vast expanse of blue sky surrounding a single, dramatic cloud formation can create a sense of awe and isolation. This approach allows the cloud to take center stage and become the sole focal point of the image.

- **Creating Minimalism:** For a more minimalist aesthetic, consider composing your image with a significant portion of the frame dedicated to negative space. This approach emphasizes the clean lines and delicate textures of the clouds.
- **Varying the Mood:** Negative space can evoke different emotions in your images. A vast expanse of blue sky with a few wispy clouds can create a feeling of peace and tranquility, while a dark, ominous cloud formation set against a minimal negative space of twilight sky can evoke a sense of drama and tension.

- **Experimenting with Balance:** Negative space doesn't necessitate a perfectly symmetrical composition. Experiment with placing the clouds off-center within the framc, maintaining a sense of balance while utilizing negative space to enhance the overall composition.

Exploring Different Angles

Don't be afraid to step outside the box and experiment with different shooting angles. Here's how varying your perspective can elevate your cloud photography:

- **Getting Low:** Get down low to the ground and shoot upwards. This perspective can emphasize towering cloud formations and create a sense of awe-inspiring grandeur. Include a foreground element like wildflowers or a winding path to add depth and context.
- **Taking Wide-Angle Shots:** Utilize a wide-angle lens to capture a vast expanse of the sky. This approach can be particularly effective during sunrise or sunset when the clouds are ablaze with color.
- **Looking Up:** Sometimes, the most compelling cloud formations are directly overhead. Lie down on your back and frame the image with interesting foreground elements like tree branches or architectural details.
- **Panoramic Compositions:** For truly expansive cloud formations, consider capturing a panoramic image by stitching together multiple shots. This technique allows you to capture the entire grandeur of the skyscape.

Using Silhouettes

Silhouettes offer a unique way to capture the drama and form of clouds. Here's how to utilize this technique effectively:

- **Shooting Towards the Sun:** The best time to capture cloud silhouettes is at sunrise or sunset when the sun is low on the horizon. Position yourself such that the sun is behind the clouds, casting their dramatic shapes against the colorful backdrop of the sky.
- **Experiment with Exposure:** For a well-defined silhouette, ensure the clouds are underexposed, appearing dark against the bright sky. Use your camera's exposure compensation to adjust the brightness and achieve the desired silhouette effect.
- **Maintaining Detail:** While silhouettes can be dramatic, aim to retain some detail within the dark clouds. Experiment with slightly adjusting exposure to find the sweet spot between a well-defined silhouette and preserving some cloud texture.
- **Adding Foreground Interest:** Silhouettes can be further enhanced by incorporating a well-defined foreground element. Imagine a lone bird perched on a branch silhouetted against a fiery sunset with dramatic cloud formations.

Remember, post-processing is a powerful tool, but use it judiciously to preserve the natural beauty of the clouds.

By mastering these composition techniques and exploring different creative approaches, you'll be well-equipped to translate the majesty of the clouds into captivating photographs. So, grab your camera, head outside, and let your creativity soar as you capture the ever-changing dance of these celestial formations.

CONCLUSION

As you embark on your journey as a cloud spotter, the vast canvas of the sky unfolds before you. You've delved into the fascinating world of cloud classification, learned to decipher their forms and messages, and honed your skills in capturing their fleeting beauty. But the journey doesn't end here.

Gazing upwards at a mesmerizing display of clouds isn't just a passive pastime; it's a gateway to a deeper appreciation of the natural world. Witnessing the ever-changing artistry of clouds evokes a sense of awe and wonder, the sheer scale prompting questions about the world around us. Cloud spotting fuels curiosity about atmospheric processes and the intricate dance between temperature, humidity, and air movement. It's a lifelong connection, accessible to anyone, anywhere. A simple glance skyward can be a source of wonder and a reminder of the beauty that surrounds us.

Cloud spotting also fosters mindfulness and stress reduction. The rhythmic movement and ever-changing patterns can have a calming effect, promoting a sense of peace and tranquility. By learning about cloud formation, we gain a deeper appreciation for the complex interconnectedness of the natural world. Clouds are not isolated entities; they are intricately linked to wind patterns, solar radiation, and the Earth's overall climate system.

The vast expanse of the sky presents a constant stream of cloud data. But filling the gaps in our understanding requires more than just sophisticated weather satellites. This is where citizen science initiatives come into play. These programs empower individuals like you to contribute valuable data by observing and reporting cloud types. Your observations, logged through apps or online platforms, provide scientists with a wealth of real-time data on cloud cover and formation across vast geographical regions.

Citizen science data contributes to research efforts in weather forecasting, climate change monitoring, and atmospheric modeling. It helps scientists validate satellite data and refine weather prediction models. Clouds play a crucial role in Earth's climate system. By observing changes in cloud cover and types over time, citizen scientists can contribute valuable data to climate change monitoring efforts. These observations can help scientists track trends and understand the potential impacts of climate change on cloud patterns.

Citizen science programs connect cloud spotters worldwide, fostering a sense of community and shared purpose. Sharing observations and collaborating with fellow enthusiasts can further enrich your cloud spotting experience. As you embark on your cloud spotting odyssey, remember that the journey is as important as the destination. Look up, marvel at the majestic formations, and become an active participant in the world of clouds. Through observation, participation in citizen science initiatives, and a continued sense of wonder, you can play a role in unraveling the secrets of the sky and fostering a deeper appreciation for our complex and magnificent planet. Let the clouds guide you on a lifelong journey of discovery, and remember, the sky is never truly empty – it's a canvas brimming with stories waiting to be told, one cloud at a time.

APPENDIX

CLOUD CLASSIFICATION TABLE

This table provides a comprehensive breakdown of cloud classification according to the international system.

Genus	Species	Varieties (if applicable)	Accessory Clouds	Supplementary Features
High Level (Cirro-)				
Cirrus (Ci)	Fibratus (Fi)	Uncinus (Unc) - Hook-shaped	Virga (Vir) - Precipitation trails	
	Uncinus (Unc)			
	Spissatus (Sp)	Patchy	Virga (Vir)	
	Nebulosus (Ne)	Sheet-like	Virga (Vir)	
Cirrocumulus (Cc)	Lacunosus (La)	Lacunaris (Lac) - Arranged in rows	Virga (Vir)	Undulatus (Und) - Rippled
	Undulatus (Und)		Virga (Vir)	Undulatus (Und)
	Floccus (Fl)	Patchy	Virga (Vir)	
Cirrostratus (Cs)	Fibratus (Fi)			
	Nebulosus (Ne)			

Middle Level (Alto-)				
Altocumulus (Ac)	Lacunosus (La)	Lacunaris (Lac) - Arranged in rows	Virga (Vir)	Castellanus (Cas) - Castellated
	Undulatus (Und)	Standing wave	Virga (Vir)	Undulatus (Und)
	Translucidus (Tr)	Thin	Virga (Vir)	Virga (Vir)
	Opacus (Op)	Thick	Virga (Vir)	Virga (Vir)
	Floccus (Fl)	Patchy	Virga (Vir)	
Altostratus (As)	Opacus (Op)		Praecipitatio (Pra) - Drizzle or light snow	Virga (Vir)
	Translucidus (Tr)	Thin	Praecipitatio (Pra) - Drizzle or light snow	Virga (Vir)
Low Level (Strato-)				
Stratocumulus (Sc)	Lacunosus (La)	Lacunaris (Lac) - Arranged in rows	Praecipitatio (Pra) - Drizzle	Virga (Vir)
	Undulatus (Und)	Standing wave	Praecipitatio (Pra) - Drizzle	Virga (Vir)
	Translucidus (Tr)	Thin	Praecipitatio (Pra) - Drizzle	Virga (Vir)

	Opacus (Op)	Thick	Praecipitatio (Pra) - Drizzle	Virga (Vir)
	Castellanus (Cas)	Castellated	Praecipitatio (Pra) - Drizzle	Virga (Vir)
	Floccus (Fl)	Patchy	Praecipitatio (Pra) - Drizzle	Virga (Vir)
Stratus (St)	Nebulosus (Ne)		Praecipitatio (Pra) - Drizzle or light snow	Virga (Vir)
Vertical Development (Nimbus-)				
Nimbostratus (Ns)			Praecipitatio (Pra) - Continuous precipitation	Virga (Vir)
Cumulus (Cu)	Congestus (Cg)	Praecipitatio (Pra) - May produce showers	Virga (Vir)	Pileus (Pil) - Cap-like
	Humilis (Hu)	Fair-weather	Virga (Vir)	
Cumulonimbus (Cb)	Incus (In)	Calvus (Cal) - Anvil-shaped top	Praecipitatio (Pra) - Heavy precipitation	Virga (Vir)
	Capillatus (Cap)	Hair-like	Praecipitatio (Pra) -	Virga (Vir)

			Heavy precipitation	
	Calvus (Cal)	Smooth Anvil	Praecip	
	Fractus (Fr)	Ragged	Praecipitatio (Pra) - Heavy precipitation	Virga (Vir) - Precipitation trails
Cloud Combinations				
Stratcumulus (Sc)	Cumuliformis (Cu for Sc)	Merging with Cumulus		Virga (Vir)
Altocumulus (Ac)	Castellanus (Cas for Ac)	Merging with Castellanus form		Virga (Vir)
Cirrocumulus (Cc)	Cirrostratus (Cs for Cc)	Merging with Cirrostratus to form patchy sheet		
Supplementary Features				
Praecipitatio (Pra)		Indicates the cloud is producing precipitation		
Virga (Vir)		Indicates precipitation trails falling from		

		the cloud but evaporating before reaching the ground		
Pileus (Pil)		Cap-like feature on top of a Cumulus cloud		
Velum (Ve)		Veil-like feature associated with a Cumulonimbus cloud		
Arcus (Arc)		Arch-shaped cloud feature associated with a thunderstorm		
Mamma (Mam)		Pouch-like protuberances on the underside of a cloud		
Praecipitatio (Pra)	Continuous	Indicates continuous precipitation		
Praecipitatio (Pra)	Intermittent	Indicates intermittent		

		precipitatio n		

GLOSSARY OF KEY TERMS

1. **Altocumulus:** Puffy white or grey patches or sheets of clouds, often arranged in rows, waves, or clusters (middle altitude).
2. **Altostratus:** Grey or bluish sheet clouds that can obscure most of the sky, sometimes with a fibrous appearance (middle altitude).
3. **Anvil Cloud:** A flat, anvil-shaped cloud associated with mature thunderstorms (any altitude).
4. **Cirrocumulus:** Thin, sheet-like clouds composed of ice crystals, often arranged in a rippled or grainy pattern (high altitude).
5. **Cirrostratus:** Thin, veil-like sheet clouds composed of ice crystals, creating a hazy or milky appearance in the sky (high altitude).
6. **Cirrus:** Wispy, feathery clouds composed of ice crystals, often appearing in streaks or patches across the sky (high altitude).
7. **Cloud Classification:** The international system for categorizing clouds based on form (appearance) and altitude.
8. **Cloud Cover:** The percentage of the sky obscured by clouds at a given time.
9. **Cloud Spotting:** The act of observing and identifying different cloud types.
10. **Convergence:** The horizontal coming together of air masses, often leading to cloud formation and precipitation.
11. **Cumulonimbus:** Towering, vertically developed clouds with a flat base, often anvil-shaped at the top, associated with thunderstorms and heavy precipitation (any altitude).
12. **Cumulus:** Puffy white clouds with flat bases and rounded tops, often appearing fair-weather clouds (low altitude).
13. **Dew Point:** The temperature at which air becomes saturated with moisture, and condensation occurs.
14. **False Cirrus:** Wispy streaks or patches that are not actually clouds but contrails (condensation trails) left behind by high-flying aircraft.
15. **Frontal Zone:** The boundary between two air masses with different temperatures and moisture content.
16. **Halo:** A luminous ring or partial ring around the sun or moon, caused by the refraction of light by ice crystals in cirrus clouds.
17. **Humidity:** The amount of water vapor present in the air.

18. **Integumentum:** A smooth, featureless sheet of cloud, often associated with altostratus or nimbostratus clouds.
19. **Inversion:** A layer in the atmosphere where temperature increases with altitude, trapping cooler air below.
20. **Lenticular Cloud:** Stationary, lens-shaped cloud that forms over mountains or other obstacles due to air flow patterns.
21. **Low Clouds:** Clouds that form at altitudes below 2 kilometers (1 mile).
22. **Middle Clouds:** Clouds that form at altitudes between 2 and 5 kilometers (1-3 miles).
23. **Mamma Clouds:** Pouch-like protuberances on the underside of altocumulus or stratocumulus clouds.
24. **Mammatus Cloud:** A cloud formation with a pouch-like or bubble-like appearance, often hanging from the underside of a larger cloud.
25. **NephOSCOPE:** A specialized instrument used to identify cloud types by measuring their angular size.
26. **Nimbostratus:** Grey featureless sheet clouds that produce continuous light rain or snow, often completely obscuring the sun (any altitude).
27. **Precipitation:** Any form of water that falls to the earth's surface, including rain, snow, sleet, or hail.
28. **Relative Humidity:** The amount of water vapor present in the air compared to the amount it could hold at a given temperature, expressed as a percentage.
29. **Stratocumulus:** Puffy white or grey patches or rolls of clouds, often arranged in a regular pattern resembling a honeycomb (low altitude).
30. **Stratus:** Grey featureless sheet clouds with a uniform base, often associated with drizzle or light fog (low altitude).
31. **Subsidence:** The downward movement of air in the atmosphere, often leading to clear skies and dry conditions.
32. **Sunshine Recorder:** An instrument that measures the amount of sunshine received at a location.
33. **Supercell:** A particularly large and powerful type of thunderstorm with a rotating updraft.
34. **Trade Winds:** Prevailing easterly winds that blow from the subtropics towards the equator.

35. **Tropopause:** The boundary between the troposphere (the lowest layer of the atmosphere) and the stratosphere.
36. **Troposphere:** The lowest layer of the atmosphere, where most weather phenomena occur.
37. **Updraft:** A rising current of air within the atmosphere.
38. **Visibility:** The furthest distance at which objects or light can be clearly distinguished.
39. **Wind Shear:** A change in wind speed or direction with altitude, which can influence cloud formation and movement.
40. **Zenith:** The point directly overhead in the sky.

Printed in Great Britain
by Amazon

44834471R00059